▼ Shiing-Shen Chern meets Chinese mathematicians at the University of California, Berkeley

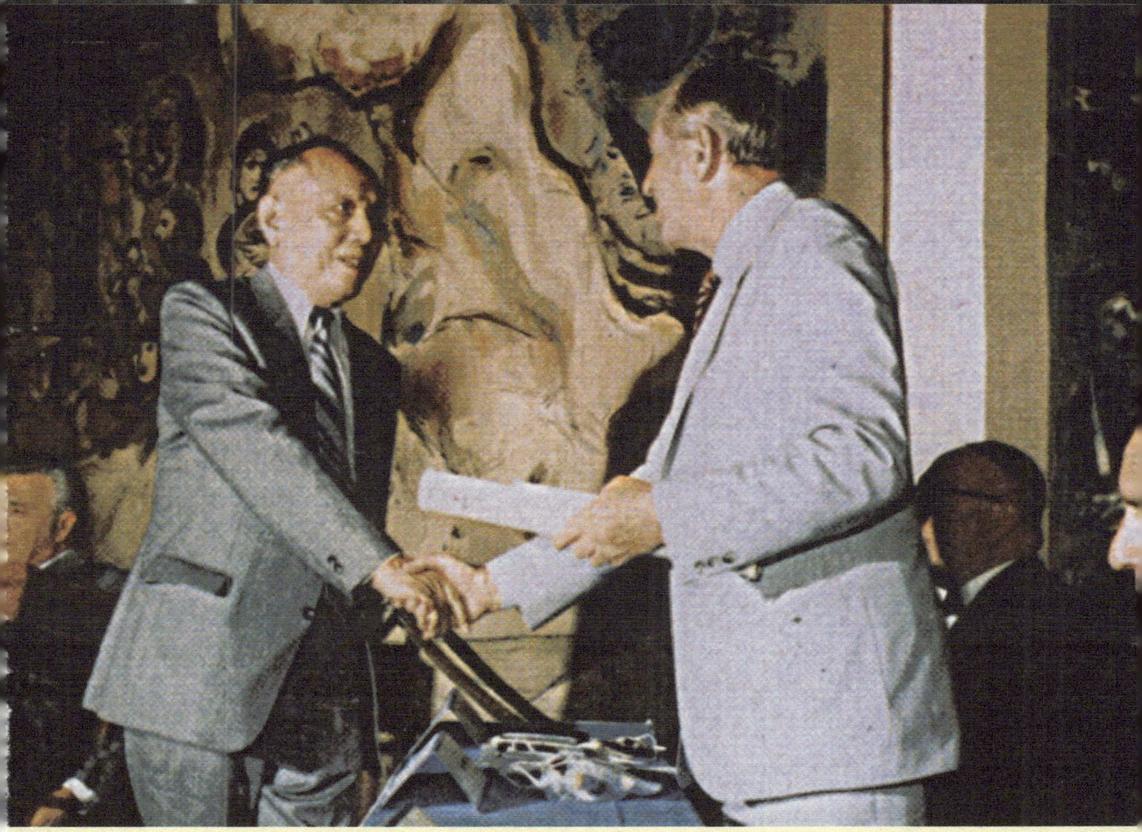

▲ Shiing-Shen Chern awarded the Wolf Prize (1984)

▲ Statue of Shiing-Shen Chern

▲ Shiing-Shen Chern

▶ Shiing-Shen Chern

How to Do Good Mathematics

www.royalcollins.com

How to Do Good Mathematics

Library *of* Mathematicians' Thoughts

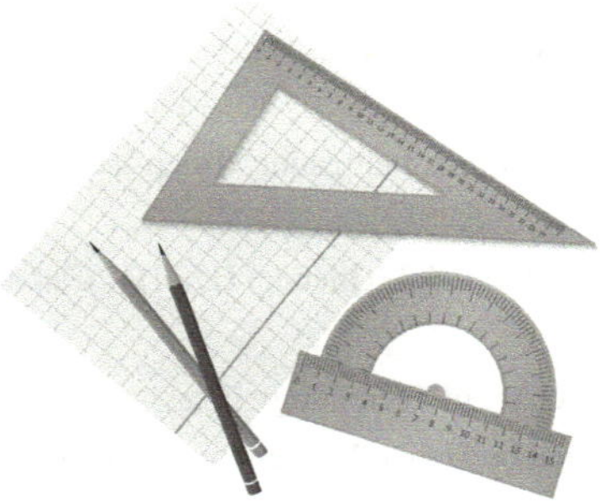

Collected Edition Editor *Li Wenlin*

Author *Shiing-Shen Chern*

Editors *Zhang Dianzhou and Wang Shanping*

Books Beyond Boundaries

ROYAL COLLINS

How to Do Good Mathematics
Library of Mathematicians' Thoughts

Collected Edition Editor: Li Wenlin
Author: Shiing-Shen Chern
Editors: Zhang Dianzhou and Wang Shanping

First published in 2025 by Royal Collins Publishing Group Inc.
Groupe Publication Royal Collins Inc.
550-555 boul. René-Lévesque O Montréal (Québec) H2Z1B1 Canada

ISBN: 978-1-4878-1344-4

To find out more about our publications, please visit www.royalcollins.com.

Contents

- Preface to the Collected Edition vii
- Reading the Masters, Approaching Mathematics ix

♦ ***Shiing-Shen Chern: Founder of Global Differential Geometry*** ***1***

- Biography and Mathematical Contributions 3
- Mathematical Thought 9

♦ ***On Mathematics*** ***23***

- Visions for the Future of Chinese Mathematics 25
- Fifty Years of World Mathematics 29
- Young Mathematicians Should Immerse Themselves in the Works of Great Masters 33
- The Mathematical Capability of the Chinese Nation: No Longer in Need of Proof 35
- Mathematics in the 21st Century 43
- Mathematics: A "Living" Discipline 53
- Do Good Mathematics: A Speech at the Shanghai Mathematical Society 55
- On the Absence of a Nobel Prize in Mathematics 61
- Infusing Humanity into Mathematics for Greater Health and Joy 63
- Mathematics: Brimming with Vitality 65

♦ ***Dialogues on Mathematics*** ***71***

◦ Mathematicians Must Pose Their Own Questions and Pursue Work with Developmental Potential: A 1991 Dialogue with Zhang Dianzou 73

◦ A Good Mathematician Should Be a Problem Solver: A Conversation with Jackson in 1998 89

◦ A Mathematician Needs Rich Imagination and Strong Resilience: A Conversation at Nankai University in 2000 99

◦ Mathematics Simplifies Science: A Conversation with *Guangming Daily* Reporter Cai Chuang 109

♦ ***Popular Mathematics Lecture*** ***113***

◦ From Triangles to Manifolds 115

◦ Characteristic Classes and Characteristic Forms 135

◦ What Is Geometry? 163

◦ Gauss-Bonnet Theorem and Maxwell's Equations 175

◦ Appendix 185

Preface to the Collected Edition

The first volume of *Library of Mathematicians' Thoughts* was released in 2009, followed by the completion of the second volume in 2021. Now, the publisher has decided to present a beautifully bound collected edition, bringing together two volumes into one exquisite compilation. Ten prominent modern mathematicians come together to share their insights on the essence, meaning, and value of mathematics, as well as to pass on the methods and spirit of mathematical innovation … The masters' thoughts are presented in their purest form.

The purpose and significance of the *Library of Mathematicians' Thoughts* has been elaborated in detail in the preface to the first and second volumes, "Reading the Masters, Approaching Mathematics," so I will refrain from repeating it here.

At this moment, China is moving toward the fulfillment of its second centennial goal. In this era of innovation-driven rejuvenation of the Chinese nation, the promotion and dissemination of scientific culture and the enhancement of the scientific literacy of the Chinese people carry immense strategic importance. We sincerely hope that the publication of this collected edition of the *Library of Mathematicians' Thoughts* will continue to shine a light on the modern endeavor of spreading mathematical culture and fostering the spirit of science.

Along with the necessary textual revisions, this collected edition includes photographs capturing the activities of the featured mathematicians. Some chapters also feature supplementary appendixes that enhance readability, infusing the serious nature of this mathematical library with a lively and engaging touch.

From the original publication of the first volume to the present collected edition, over a decade has passed. During this time, the dedication of the editors and translators, the perseverance of the publisher, and the invaluable support and feedback from our readers have been indispensable. As I stand before the proofs of the ten-volume collected edition now ready for release, I share in the satisfaction and reflection of the editors, while also feeling that our work is far from complete. We will continue to forge ahead!

Li Wenlin

November 2022

Zhongguancun, Beijing

Reading the Masters, Approaching Mathematics

Mathematical Thought: The Soul of Mathematicians

Mathematical thought is the soul of mathematicians. Imagine this: Without axiomatic thinking, could we ever have Euclid or Hilbert? Without the fusion of numbers and geometry, where would Descartes be? Without the concept of mathematical structure, how could we even begin to speak of the Bourbaki school?

The essence of a mathematician's thought is primarily embodied in their groundbreaking mathematical research—through the new concepts, theories, and methods they introduce. Newton and Leibniz's calculus, Gauss, Riemann, and Lobachevsky's non-Euclidean geometries, Galois' notion of "groups," Gödel's incompleteness theorems and Turing machines, Nash equilibrium theory, and so on—all come together to form an awe-inspiring ocean of mathematical thought, shaping the indelible chapters of human intellectual history.

But the mathematical philosophy of these mathematician is also part of mathematical thought. Their reflections on the nature, feature, significance, and value of mathematics, their views on the sources of mathematical knowledge, and its connection with other fields of human inquiry, as well as their insights into scientific methodology—these too are integral components of their mathematical thought. Of course, throughout history, mathematicians have often held differing, sometimes opposing views on these issues. Yet, it is precisely these diverse voices that converge to create a symphony of rational thought.

Just as one comes to understand and appreciate an artist or a composer through their paintings or compositions, mathematical thought is undoubtedly the primary way in which we come to know and assess mathematicians. It is also the core of their contribution to humanity and the very substance from which people seek knowledge. In this light, we might say: "Mathematicians think, and thus mathematicians exist."

The Societal Significance of Mathematical Thought

Is mathematical thought something only mathematicians should possess? Certainly not. Mathematics is the foundation of natural sciences, technology, and the humanities—a fact that has become increasingly recognized as a cornerstone of modern society. This foundational role of mathematics is not only due to its use as the language and tool of science, finding applications in nearly every field of human knowledge, but perhaps even more so for its profound cultural function: to nurture and develop the human mind, especially its capacity for precise thinking. Regardless of the career one pursues, one's thinking ability is an invaluable, invisible asset, and mathematics is the perfect discipline to hone this ability. This is why mathematics remains one of the subjects that demands the most time throughout a person's education.

This, however, does not mean that every specific mathematical concept learned in school will have direct, practical use later in life. The true impact of mathematics on lifelong development lies in its transformative power over one's way of thinking. Take Euclidean geometry, for example. While most of the theorems we study in school may rarely find direct application in our daily lives or professional work, the rigorous deductive reasoning and logical thinking inherent in Euclidean geometry play an undeniable role in shaping the elite minds of every field. In fact, from Newton's *Mathematical Principles of Natural Philosophy* to Einstein's writings on relativity, from the Declaration of the Rights of Man and of the Citizen during the French Revolution to Marx's *Capital*, and even in the works of modern Nobel laureates in Economics, one can trace the influence of Euclidean thought. Moreover, the quantitative thinking embedded in mathematics has, with remarkable breadth and depth, permeated nearly all areas of human knowledge. From its rigorous proofs to its precise calculations, mathematics offers humanity a model of exact and refined thinking.

A striking example can be found in the concept of "program memory" or the idea of "program automation," which plays a pivotal role in modern computer design. When the first electronic computer, ENIAC, was initially constructed, it faced the imminent risk of failure due to the sharp conflict between its increasing computational speed and the sluggishness of manual programming. At this critical juncture, it was the brilliant mind of mathematician John von Neumann who introduced the concept of "program memory," ultimately rescuing this groundbreaking technological invention. Even today, the fundamental principles of computer architecture continue to adhere to von Neumann's visionary ideas. For this reason, he is rightly revered as the "father of the computer" (although it is now recognized that he was not the sole mathematician to propose such an idea). Isn't it fascinating that a concept like "program memory," which at first glance might not seem "mathematical," was created by a mathematician—by a genius mathematician like von Neumann?

Thus, we can assert that the mathematical thought of mathematicians is a priceless treasure for society as a whole. The propagation of mathematics, beyond the transmission of specific mathematical knowledge, lies at the heart of spreading mathematical thought itself. In today's world, where science and technology are increasingly shaped by mathematics, this has become an urgent societal necessity. Just imagine: If more and more individuals could harness mathematical thinking to think and solve problems, how profoundly that could shape the progress of society!

Reading the Masters, Approaching Mathematics

Mathematics is the art of numbers and shapes, and the creative minds of mathematicians are as vibrant as they are original. Their thinking is neither bound by convention nor can it be reduced to rigid dogma. Understanding a mathematician's thought can be approached in many ways, but reading the original works of mathematicians—especially the great masters—is perhaps the most direct, reliable, and rewarding path.

Mathematicians' writings can generally be categorized into two types. The majority is their technical monographs, where they discuss their theories and methods. For those engaged in truly original research, these foundational works are the most vivid and instructive of texts. As Laplace often advised his students: "Read Euler, read him—he is the teacher of us all." Here, Laplace was speaking primarily to professional mathematicians and physicists; for the general public, such works are often difficult to approach.

The other category of works is more accessible, often written for a broader audience. These include reflections on the philosophy of mathematics—what Godfrey Harold Hardy referred to as "mathematics about mathematics"—as well as more popular expositions of mathematical knowledge and the mathematician's own creative contributions. Unlike the obscure technical monographs, these writings are highly readable, more easily embraced by the public. Many of these works are not only intellectually enriching but also artistically elegant. Indeed, several great mathematicians were also masters of language, and if writing is considered an art of language, then these works exemplify the union of mathematics and art. Reading these masterpieces is an aesthetic experience, offering a way to learn about mathematics while savoring the beauty of the language, immersing oneself in the culture and allure of mathematics. This is the very purpose of compiling *Library of Mathematicians' Thoughts*.

Library of Mathematicians' Thoughts selects representative works by some of the most renowned modern mathematicians, with one chapter dedicated to each mathematician. The translation went one after another, and was compiled into two volumes for publication. The first volume offers a curated collection of writings from 20th-century mathematical masters such as D. Hilbert (1862–1943), G. Hardy (1877–1947), J. von Neumann (1903–1957), Bourbaki (1935–present), M. F. Atiyah (1929–2019), etc. (the latter three being reprints). It has been warmly received and reprinted multiple times. Encouraged by this success, we continued with the second volume, which features the writings of mathematicians such as F. Klein (1849–1925), H. Weyl (1885–1955), A. N. Kolmogorov (1903–1987), Hua Luogeng (1910–1985), and Shiing-Shen Chern (1911–2004). These works, often short but brimming with profound insight, have circulated widely, both in China and abroad. They are, in essence, like precious gems in the vast ocean of mathematical thought or beautiful blossoms in the garden of mathematics.

We do not expect these gems or blossoms to reverse the tide of utilitarianism, but we believe in the words of Einstein, who, in honoring Newton, said: "**The products of understanding outlast the noise of passing generations. They endure for centuries, continuing to radiate light and heat**."

We sincerely hope that the "products of understanding" of the great mathematicians featured in this collected edition will continue to shine brightly, spreading the light of mathematical thought and advancing the cause of scientific culture.

Reading the masters and approaching mathematics—everyone stands to benefit.

Li Wenlin

Researcher, Institute of Mathematics and Systems Science

Chinese Academy of Sciences

July 2021

Zhongguancun, Beijing

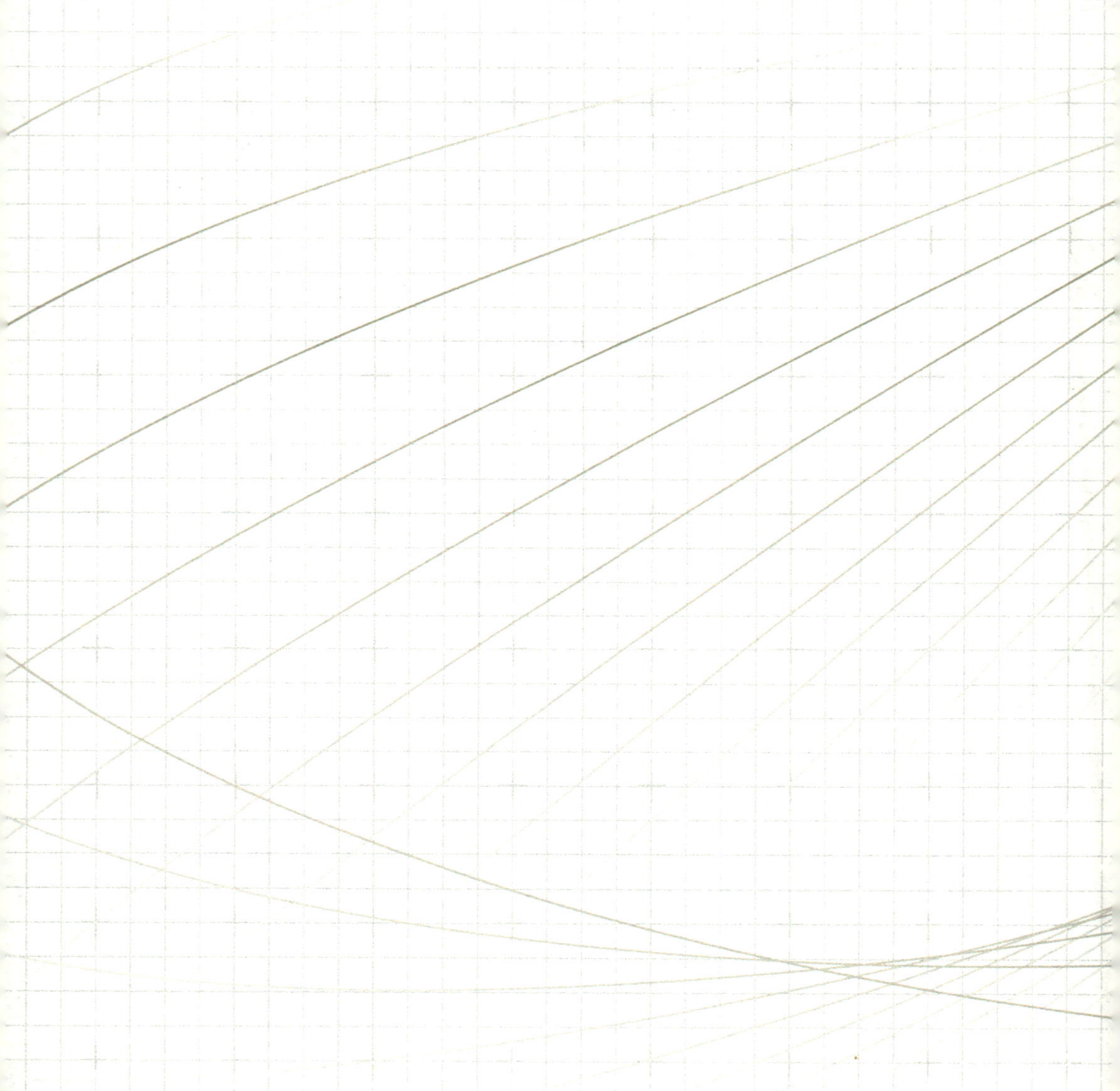

Shiing-Shen Chern: Founder of Global Differential Geometry

Biography and Mathematical Contributions

Early Life and Education

Shiing-Shen Chern was born on October 28, 1911 (the 7th day of the 9th lunar month) in Xiatang Street, Jiaxing, Zhejiang Province (now located at No. 665, Jianguo Road, Jiaxing City, which was then part of Jiaxing Prefecture's Xiushui County). His father, Chen Baozhen (1889–1967), was a *xiucai* (秀才, scholar who has passed the entry-level examination to study at a college) during the late Qing Dynasty (1616–1911) and entered the judicial system following the 1911 Revolution. In 1923, he was appointed to a position at the Tianjin Court and moved his family there.

Shiing-Shen Chern did not attend private schools or elementary schools. In 1920, he enrolled in Xiuzhou Middle School in Zhejiang, where he began his preparatory studies. After relocating with his family to Tianjin in 1923, he transferred to Tianjin Fulun High School, where he completed the second semester of the first year. Upon graduating from Fulun High School in July 1926, he entered the Faculty of Sciences at Nankai University in September, focusing on mathematics from his second year onward. Under the guidance of Professor Jiang Lifu (1890–1978), he developed a deep interest in geometry.

In June 1930, Chern graduated with top honors from Nankai University and, in September, entered graduate school at Tsinghua University. There, he studied under Dr. Sun Guangyuan (1900–1979) and specialized in projective differential geometry. In 1932, the esteemed German geometer W. Blaschke (1885–1962) visited China and gave a series of lectures at Peking University on the Topological Problems in Differential Geometry. Chen attended every lecture, and Blaschke's clear and insightful discussions opened his eyes to the fact that projective geometry, while important, was a peripheral branch of mathematics that had drifted from the main currents of mathematical development.

In 1934, Chern earned his master's degree with a thesis on projective line geometry. His outstanding academic performance earned him a scholarship to study abroad. Though the funding was tied to the return of the Boxer Indemnity, which traditionally required students to study in the United States, he applied instead to study under Blaschke at the University of Hamburg in Germany and was granted approval.

In September 1934, Chern arrived at the University of Hamburg, where he studied network geometry under Blaschke. In just one year, he completed his doctoral dissertation and successfully defended it in February 1936, earning his PhD. As a result of his early graduation, Chern received additional funding from the China Cultural Foundation for another year of study. He chose to spend this year at the University of Paris, studying under Elie Cartan, one of the most prominent geometers of the early 20th century. Although Cartan's work was notoriously difficult, Chern immersed himself in it, making rapid progress and refining his own mathematical understanding.

By the time Chern was in Paris, he had already been appointed as a professor at Tsinghua University. In the summer of 1937, the outbreak of the Chinese People's War of Resistance against Japanese Aggression forced Tsinghua University, along with Peking University and Nankai University, to relocate to Changsha, where they merged to form the Temporary University of Changsha. Chern made his way to Changsha via Hong Kong and began teaching there. In February 1938, the university moved once more, this time to Kunming, Yunnan, and was renamed the National Southwestern Associated University. Chern followed the university to Kunming, where, in April 1939, he married Miss Shih-Ning Cheng. Despite the difficulties of wartime life, he continued his teaching while devoting himself to studying Cartan's works, through which he was able to grasp the core of Cartan's geometric theories.

A New Era in (Global) Differential Geometry

In August 1943, Chern was invited to visit the Institute for Advanced Study in Princeton, US. Three months later, he wrote the groundbreaking paper "A Simple

Intrinsic Proof of the Gauss-Bonnet Formula for Closed Riemannian Manifolds," which would come to be regarded as one of the most important mathematical works of his career.

As is well-known, the formula stating that "the sum of the interior angles of a triangle in a plane equals 180°" was established over two thousand years ago by the ancient Greek scholar Euclid (active around 300 BC) and formed the foundation of classical geometry. In the 19th century, the great German mathematician Carl Friedrich Gauss (1777–1855) pioneered differential geometry and extended this formula to triangles on curved surfaces. Later, the French mathematician P. O. Bonnet (1819–1892) generalized it further to regions enclosed by closed curves.

Manifolds, the generalization of Euclidean space, are locally equivalent to Euclidean space but can exhibit a far more intricate structure globally. The central challenge of early 20th-century differential geometry was extending the Gauss-Bonnet formula to these more complex manifolds. This was especially difficult for high-dimensional manifolds with complex structures, and at the time, mathematicians were unsure how to express the formula for them. The renowned geometer H. Hopf (1894–1971) once remarked, "Extending the Gauss-Bonnet formula to high-dimensional compact manifolds is an extremely important and difficult problem in geometry."

In his concise six-page paper, Chern introduced a revolutionary approach using Elie Cartan's exterior differential forms and moving frames. By performing operations on the unit-length tangent bundle and integrating topological theory, he succeeded in proving the Gauss-Bonnet formula for general closed Riemannian manifolds (in even dimensions). This breakthrough illuminated the path forward for the field, and terms such as "exterior differential forms," "connections," "frames," and "fiber bundles" quickly became central concepts in differential geometry.

In September 1945, Chern was invited to deliver a one-hour lecture at the American Mathematical Society's Summer Meeting, titled "Some New Viewpoints in Differential Geometry in the Large." The talk had a profound impact on the mathematical community, with H. Hopf noting, "This lecture signals the beginning of a new era in global differential geometry, characterized by the synthesis of fiber bundle topology and Cartan's methods."

In October 1945, Chern completed another major paper, "Characteristic Classes of Hermitian Manifolds," in which he introduced a new invariant for characterizing the structure of manifolds. This would later be known as the "Chern class." The

Chern class has widespread applications in modern mathematics, particularly in topology, geometry, complex analysis, and algebraic geometry. It has even been deeply applied in theoretical physics.

Founding and Leading the Institute of Mathematics at the Academia Sinica

In early April 1946, Chern returned to Shanghai, China. Shortly after, he was appointed by the Academia Sinica to establish an Institute of Mathematics, succeeding Professor Jiang Lifu, who had gone to America for a visit. Once the institute was formally established, Chern was appointed as its acting director and began overseeing its operations.

Chern viewed the cultivation of young mathematicians as his foremost responsibility. He actively recruited the most promising students from mathematics departments across China's universities to join the institute, personally teaching a demanding 12-hour weekly course in topology. In less than three years, he had mentored a group of outstanding young mathematicians, including Liao Shantao, Chen Guocai, Yang Zhongdao, Wu Wenjun, Zhang Sucheng, Zhou Yulin, Lu Jianke, Cao Xihua, Ye Yanqian, Chen Jie, Chen Dehuang, Lin Xian, He Xizhang, Ma Liang, and Sun Yifeng. Many of these individuals would go on to become the backbone of Chinese mathematics, contributing greatly to Chinese mathematical development.

Leading the Revival of Geometry in the United States

In January 1949, Chern returned to the United States with his family, resuming his role as the principal speaker for the "Veblen Seminar" at the Institute for Advanced Study in Princeton. Later that summer, he accepted an invitation to join the University of Chicago as a professor in the Department of Mathematics, where he would mentor ten PhD students over the course of the next eleven years.

In 1950, Chern delivered a one-hour lecture at the 11th International Congress of Mathematicians, titled "Differential Geometry of Fiber Bundles." In 1958, he returned to the Congress to give a 30-minute talk on "Differential Geometry and Integral Geometry." In June 1960, he left the University of Chicago and accepted a position at the Department of Mathematics of the University of California, Berkeley, where he would spend the next two decades. During this time, he transformed the department into a global center for geometry and topology, nurturing 31 PhD students. In 1970, he gave another hour-long lecture at the 16th International Congress of Mathematicians on "The Past and Future of Differential Geometry."

Chern retired from UC Berkeley in 1979, taking the title of professor emeritus. Two years later, in 1981, he became the founding director of the Mathematical Sciences Research Institute.

When Chern first arrived in the US, the field of differential geometry was in sharp decline, with some even declaring "it is dead." Under his guidance, however, American geometry experienced a renaissance, emerging as a leader in global research. The American geometer R. Osserman (1926–2011) would later note, in an article celebrating the 100th anniversary of the American Mathematical Society, that "but for a single most decisive factor contributing to the rebirth of geometry in America, I would propose the move of Chern to the United States from China in the late forties."[1]

Founding the NKIM and Devoting His Life to China's Mathematical Development

In September 1972, Chern, accompanied by his wife Shih-Ning Cheng, visited China for the first time in 24 years. He delivered a lecture on "Fiber Spaces and Characteristic Classes" at the Institute of Mathematics, Chinese Academy of Sciences. From that point on, Chern made nearly annual visits to China, where he frequently expressed his desire to dedicate his remaining years to advancing mathematics in his homeland when talking to old friends.

---

1 R. Osserman, *The Geometry Renaissance in America: 1938–1988*. (For Chinese translation, see reference [2], pp. 387–393.)

In 1981, Chern met with Hu Guoding three times, then Vice President of Nankai University, for a series of discussions about establishing the Nankai Institute for Mathematics (NKIM). Over the next few years, Chern dedicated a significant amount of time and energy to bringing this vision to life, securing support from a variety of sources, and making substantial personal contributions. On October 17, 1985, the NKIM was officially founded, with Chern as its first director. It hosted the "Nankai Mathematics Year" event for 12 consecutive years. These annual gatherings, which invited international experts to lecture for young mathematicians and graduate students from across the country, became a central event in China's mathematical community.

On January 12, 2000, Chern's beloved wife, Shih-Ning Cheng, passed away. Amid his grief, He expressed a poignant wish to the authorities: when their time came, the couple's ashes would be laid to rest at the Nankai campus, without a tombstone or a monument, but beside a blackboard. This board, he said, would serve as a place for future generations of students to continue their mathematical studies.

In September 2000, Chern officially returned to China to settle down permanently.

At Chern's suggestion, along with that of fellow mathematician Shing-tung Yau, China successfully bid to host the International Congress of Mathematicians. In August 2002, the Congress was held in Beijing's Great Hall of the People, with Chern elected as the honorary chair of the event.

On December 3, 2004, at 7:14 p.m., Chern passed away from a myocardial infarction at Tianjin General Hospital. In accordance with his and his wife's wishes, their ashes were interred at Nankai University. The design of their final gravesite is a tribute to a mathematics classroom: Around the site, approximately twenty wooden benches invite future scholars to pause, reflect, and engage in mathematical discussions. At one corner stands a black monument, akin to a blackboard, inscribed with Chern's mathematical manuscripts, including the renowned Gauss-Bonnet formula.

Mathematical Thought*

Do Good Mathematics

Shiing-Shen Chern frequently pointed out that for China to become a "mathematically advanced country," it must focus on doing good mathematics. It is only through the creation of truly good mathematics that the country can cultivate its own distinctive voice and earn an "independent and equal" standing within the international mathematical community.

In 1992, during the Celebratory Symposium of the Tenth Anniversary of the National Natural Science Foundation of China, Chern revisited this theme. He remarked as follows:

> A mathematician should understand what constitutes good mathematics and what is bad or less refined mathematics. Some mathematics is groundbreaking and has promising development potential—that is good mathematics. There is also mathematics that is quite interesting but gradually turns into a mere game.
>
> Let me illustrate this with an example. Perhaps you are familiar with Napoleon's Theorem. It is said to have some connection to Napoleon himself. The theorem states that for any triangle, if equilateral triangles are constructed on its sides, the centroids of these triangles will form an equilateral triangle. If the equilateral triangles are constructed inward instead, two solutions emerge. This type of mathematics is not considered "good" because it offers little opportunity for further development. Of course, if

* The excerpts from Chern's speeches and the examples cited can be found either in this book or in the references provided in the bibliography.—Ed.

you're looking to unwind and want to ponder on this question casually while lounging on the couch, this could be enjoyable—almost like a mental game for relaxation.

So, what does "good" mathematics look like? Take solving equations, for example. Solving equations is the cornerstone of mathematical inquiry. Solving linear equations is easy, but quadratic equations are different. For example, the equation $x^2 - 1 = 0$ has real solutions, while $x^2 + 1 = 0$ does not. Eventually, the introduction of complex numbers allows us to explore the complex solutions to equations. The famous Fundamental Theorem of Algebra asserts that every polynomial equation of degree n has a solution in the complex number system. This issue has a long and storied history. The great mathematician Euler (1707–1783) pondered this problem, but it wasn't until Gauss (1777–1855) proved it that we began to understand the connection between complex numbers and topology. This breakthrough led to an entirely new realm of insight. A complex number with modulus 1 represents a point on a circle, and there are infinite patterns on that circle. When we move from solving $f(x) = 0$ to $f(x,y) = 0$, we enter the study of curves, which may or may not have solutions—there may not even be a real solution at all. And that's when the variation unfolds. If we treat both x and y as complex numbers in equation $f(x,y) = 0$, the two complex numbers correspond to a four-dimensional space, leading us into the intricate world of Riemann surfaces in complex analysis. Solving the original equation $f(x,y) = 0$ through the lens of Riemann surfaces requires a wealth of mathematical knowledge, including the crucial concept of (genus) g. When we interpret the solutions to $f(x,y) = 0$ as a curved surface, the number of circles on the curved surface is linked to the (genus) g, which distinguishes different variations of surfaces.

Furthermore, we can explore even more fascinating variations. For example, if we assume the coefficients of $f(x,y) = 0$ to be integers, we can look for its integer solutions. This adds another layer of complexity.

This example clearly illustrates what we mean by "good" mathematics—those problems that have profound and lasting significance, that can be explored in depth, and that impact a multitude of fields. Equations, for instance, are perennial problems with eternal value, continually evolving, which is why they are quintessentially "good" mathematics.

In his report in Shanghai Mathematical Society on January 6, 1994, Chern reiterated his views on what makes "good" mathematics. He cited the 18th-century French mathematician Lagrange (1736–1813), who set forth two criteria for identifying valuable mathematical problems: First, they must be understandable—one should be able to explain them to anyone, anywhere. Second, they must be challenging—difficult, yet not impossible to tackle. Problems that meet both of these conditions are rare. In 1900, the German mathematician Hilbert (1862–1943) presented 23 problems, all of which were exemplary: challenging yet accessible, and they played a pivotal role in shaping the course of 20th-century mathematics. In his report, Chern also pointed out that Fermat's Last Theorem, $x^n + y^n = z^n$ $(n > 2)$ has no integer solutions. The meaning of the problem is easy to understand, but solving it is extremely difficult. Similarly, the three-body problem, which studies the orbital equations of the sun, earth, and moon, is also easy to understand but incredibly complex to solve. Fermat's Last Theorem was eventually proved by Andrew Wiles in 1995, and progress continues in the study of the three-body problem.

In his report in Shanghai, Chern also addressed the nature of problems in the Math Olympiad for middle school students. He said, "I support math competitions and always encourage the winners, wishing them success. However, the problems posed in these competitions are not truly significant. Problems that young students can solve within two or three hours cannot possess profound meaning. This is not to say that Olympiad problems are inherently bad, but rather that winning such a competition demonstrates a student's ability, not their capacity for researching a meaningful mathematical problem. It is also a mistake to equate Math Olympiad winners with professional mathematicians."

"The best mathematics should offer new perspectives. Merely copying others' work is absolutely not good mathematics. In science, most experiments and research efforts end up being wasted, with only a few that truly stand the test of time. As mathematicians, we believe that if mathematics is truly good, it will undoubtedly have applications."

So, what is the relationship between good mathematics and current mainstream mathematics? Chern once explained, "Mainstream mathematics refers to significant contributions, profound theorems that have broad implications and complex proofs. If I were to choose an example of such a contribution today, it would be the Atiyah-Singer Index Theorem. This theorem represents the modern extension of the classical $f(x,y) = 0$ problem, generalizing concepts like algebraic equations, Riemann surfaces, and genus from low dimensions to higher and even infinite dimensions."

Mainstream mathematics is undoubtedly important, and research and study in this field is crucial. However, Chern believed that there are many directions in mathematics, and since it is a personal pursuit, not everyone must work within the mainstream. "I think we should encourage people not to confine themselves to mainstream mathematics. The ideal scenario is that what is not considered mainstream today may become mainstream in the future." He often used his own experience to highlight this. Before World War II, differential geometry was not seen as mainstream and was even considered "dead." Yet, by the second half of the 20th century, it had become a central field of study. Chern, having been a pioneer in the field, helped lay the foundation for global differential geometry.

Chern also observed that smaller countries often excel in specialized fields. For example, in the early 20th century, Poland focused on set theory and functional analysis, while Finland concentrated on the value distribution of complex functions, producing some of the world's leading experts in the field. These countries, with their unique feature and strength, have made significant contributions to world mathematics. Doing good mathematics requires skill, just as finding oil in a vast desert requires the ability to identify where to drill. Randomly digging wells will not work. Mathematics is no different. Those who are capable of doing good mathematics are those who are hardworking, though hard work can take many forms. For instance, van der Waerden (1903–1996) from Leipzig, Germany was known for working diligently at his desk all day, while S. Cohn-Vossen (1902–1936) often wandered, thinking about mathematics as he walked. Yet both were accomplished mathematicians.

Chern also noticed a growing trend in the choices of talented students today. Many are turning to fields such as computer science, economics, and management. He believed that in the years to come, there would be a shortage of mathematical talent. Moreover, mathematics is not a field that anyone can easily enter; becoming a mathematician requires years—often decades—of dedicated training.

The Enchanting Applications of Abstract Mathematics

Chern once remarked, "Mathematics is a curious thing. It may appear extremely abstract, as though many concepts are mere constructs of our minds, abstract problems with no immediate real-world connection. However, when we look back at the history of mathematics over the past thousands of years, we see that these abstract ideas have proven to be profoundly useful. Many of these abstract results have far-reaching applications in diverse fields."[1]

He offered a few examples to illustrate this. For instance, the theory of regular polyhedra from classical geometry has been applied by chemists to study molecular structures. Knot theory in topology has become an invaluable tool for biologists researching DNA structures. Even the mathematical constant pi (π) appears frequently in demographic studies.

Perhaps the most extraordinary application of abstract mathematics in the 20th century was Chern's creation of global differential geometry in the 1940s. Astonishingly, thirty years later, this field found a deep connection with theoretical physics!

▸ Shiing-Shen Chern and Chen-Ning Yang: Meeting the Same Elephant from Different Angles

In 1954, two young physicists, Chen-Ning Yang (1922–) and Mills (1927–1999), developed a groundbreaking theory to address the problem of isospin symmetry—non-Abelian gauge theory. This theory would later become the cornerstone of modern theoretical physics, explaining the interactions between electromagnetism, weak force, and strong force.

Then, in 1975, Yang made a stunning realization: gauge fields were, in fact, the same as the connections on fiber bundles in differential geometry! The discovery struck him like the force of an electric shock. He later wrote, "The mysteries of the objective universe aligned perfectly with mathematical concepts developed through pure logic and the pursuit of beauty. It was truly staggering." That evening, Yang

---

1 Shiing-Shen Chern, "The Enchanting Applications of Abstract Mathematics." See reference [2], p. 140.

drove to Chern's home, sharing that he had understood the elegant theory of fiber bundles and the profound Chen-Weil Theorem. "The gauge is exactly the connection on a fiber bundle ... I was astonished and puzzled, because you mathematicians somehow created these concepts out of thin air."

Chern immediately disagreed, saying, "No, no! These concepts weren't imagined. They are natural and real." Reflecting on the conversation later, Chern remarked: "Yang-Mills gauge theory was published in 1954, and my work on characteristic classes in 1946. We had no idea our works were so closely related. It wasn't until 20 years later, as both became better understood, that we realized we had been encountering different parts of the same elephant."[2]

Just as Einstein's general theory of relativity revolutionized Riemannian geometry, Yang-Mills gauge theory propelled differential geometry to the cutting edge of modern science. What was once considered an abstract creation in the minds of mathematicians quickly became a very real, very tangible physical structure. This discovery sent ripples through the scientific community in the 1970s, attracting the attention of both physicists and mathematicians alike.

Another remarkable event occurred in 1974 when Chern, in collaboration with American mathematician J. H. Simons (1938–), introduced a new geometric invariant, later known as the Chern-Simons invariant. No one could have predicted that this purely mathematical concept would eventually find a profound connection with physics, particularly with the Yang-Mills gauge theory. This led to the creation of an entirely new field of physics—topological quantum field theory. Today, the Chern-Simons invariant is one of the most frequently cited terms in both mathematics and physics literature.

▸ Physics Is Geometry

On September 24, 1999, Chern delivered a lecture at Fudan University titled "What Is Geometry?" In his speech, he eloquently explored the profound relationship between physics and geometry.

2 Shiing-Shen Chern, "My Connection with the Yang Family." See reference [2], pp. 78–80.

I'm sitting next to two great physicists (referring to Chen-Ning Yang and Xie Xide). Now, I'd like to share my thoughts on the relationship between physics and geometry. I believe that much of the essential work in physics revolves around proving that physics is fundamentally geometry. Take Newton's second law of motion, for example. Newton's second law states: $F = ma$. Here, F is force, m is mass, and a is acceleration. We now call acceleration curvature. Thus, the right-hand side of the equation is a geometric quantity, while force is a physical quantity. In essence, Newton spent all that time simply stating that physics is geometry! (Laughter and applause)

And the same is true with Einstein's general theory of relativity. The equations of general relativity are $R_{ik} - (1/2)\, g_{ik}\, R = 8\pi K T_{ik}$. Here, R_{ik} is the Ricci curvature, R is the scalar curvature, K is a constant, and T_{ik} is the energy-stress tensor. If you carefully examine this equation, you'll see that the left-hand side consists of geometric quantities derived from the Riemannian metric. So, Einstein's key equation essentially says that geometric quantities equal physical quantities. (Applause)

The Absence of a Nobel Prize in Mathematics: A Blessing

The Nobel Prize, renowned as the first global scientific honor to disregard nationality, race, gender, and ideology, has come to symbolize the pinnacle of academic achievement. It is awarded solely based on scholarly merit, making it the most coveted accolade for any scientist. Each year, the announcement of the laureates creates a media frenzy. Winners bask in immense glory, with their institutions and even their countries often sharing in the honor. The number of Nobel laureates has even become a barometer for assessing a nation's scientific prowess. However, mathematics is conspicuously absent from the Nobel Prize categories—a fact that has stirred discontent in some and regret in others.

Chern, however, held a rather unique viewpoint. "In the grand scheme of things, I believe that the absence of a Nobel Prize for mathematics is, in fact, a blessing for the field." He elaborated, "Perhaps it's better that mathematics doesn't have a Nobel Prize. The attention such a prize would bring might distract mathematicians from focusing on their research." He continued, "Mathematics is filled with simple yet profound problems. Delving into these challenges brings immense joy. It is a great discipline—its growth intersects with many other sciences, and it stands as a true marvel of human thought. Mathematics is characterized by its many deceptively simple yet deeply difficult problems. These problems can consume your thoughts and lead to sleepless nights, or remain unsolved for years, but the joy that comes from solving them is beyond words."

Chern's appreciation for "the quiet world of mathematics" remains a shared haven for mathematicians to this day. A case in point is Andrew Wiles, who, after retreating into seclusion at Princeton for eight years, finally cracked Fermat's Last Theorem. He spent those years immersed in the quiet world of mathematics, quietly dedicating invaluable time to the problem.

Chern was not critical of the lack of a Nobel Prize in mathematics; rather, he found the reasoning behind it simple and sensible.

> The Nobel Prize honors contributions that directly impact human happiness. That's why it includes categories like Peace, Medicine, and Literature. The founders were farsighted enough to foresee the massive developments in fields like Physics and Chemistry—this was truly remarkable foresight. The first prize was awarded in 1901, and the first Physics prize went to Wilhelm Roentgen for his discovery of X-rays.
>
> Mathematics, however, serves an indirect role. Without complex numbers, there would be no electromagnetism; without Riemannian geometry, no general relativity; without the geometry of fiber bundles, no gauge theory … The profound study of material phenomena is deeply intertwined with advanced mathematics, which remains a mysterious phenomenon in scholarship.
>
> Science requires experimentation, but experiments can never be absolutely precise. In contrast, if there is a mathematical theory, it can provide entirely correct inferences. This is why science cannot do without mathematics. Many fundamental concepts in science are expressed through mathematical frameworks. Hence, mathematicians continue to make a living, though it's only natural that they do not receive a Nobel Prize.

As the number of global awards increases, mathematics does have its own prestigious honors, such as the Fields Medal and the Wolf Prize. Chern himself was awarded the Wolf Prize in 1984. He believed that winning such an award was a form of social recognition to be cherished, but it should never be flaunted. We must respect others who receive awards, but we should not idolize them. An award is a result, not a goal.

In 1999, during the presentation of the Qiu Shi Science Prize, Chern shared his thoughts with the young audience.

> Today, we often talk about awards. When we present an award, with its prize money, it's a form of social and governmental acknowledgment of your work. In the past, in Europe, if you didn't have a position as a mathematics professor, no one would pay you a salary. One major way to earn a living was through prizes. Several academies offered prize money, which could sustain you for a while, and that was a great joy. Interestingly, however, when Riemann discovered the Riemann-Christoffel curvature tensor, he applied to the French Academy of Sciences for prize money. The academicians didn't understand his work, so they didn't offer him the grant. So, to those of you sitting here today in the front rows, all of you are prize winners, honored by society. There are great expectations for you. But many in the back rows have never won an award. Don't worry—this doesn't matter. You can still do meaningful work. I, too, did not win an award before I received my degree. Winning an award is not the most important thing. Even Riemann didn't win an award. His Riemann tensor, when he applied for a prize at the French Academy, was not recognized.

In conclusion, awards are merely the result, not the goal. Rather than chasing, competing, or scrambling for prizes, it's better not to seek them at all.

Focusing on Mathematics Education

▸ Math Is Fun

In August 2002, Chern wrote the phrase "Math Is Fun" for the China Youth Mathematics Forum. This simple yet profound statement had never been used before in the history of Chinese mathematics education, and it struck a deep chord with teachers and parents across the country. In a matter of days, these words spread like wildfire, creating a fresh and powerful wave of inspiration that captured the hearts and minds of many. For Chinese children, who often carried heavy academic burdens, this message was both timely and crucial, opening up a whole new world of possibilities in their education.

Chern explained: "Mathematics is a fascinating science. That's why I wrote 'Math Is fun' for the children. Mathematics lessons should be taught in a way that sparks their interest. Children are inherently curious. They naturally have an interest in mathematics. However, if it's taught in a dry, uninspiring manner, it stifles their curiosity and makes mathematics feel difficult."[3]

He further added: "We should not measure everyone with the same ideal yardstick. The old Chinese teaching principle is teaching according to the student's ability. Today, however, Chinese education places too much emphasis on scores, and everyone is judged by a single total score. This is not unlike the old imperial examination system, which used rigid essay-writing standards to select talent. This approach is not entirely rational. We should allow more flexibility in how we select talent, letting children express their abilities freely. This should be our educational goal."

His vision for Chinese mathematics education was as follows: "Follow our own path and avoid blindly imitating American math education. Our students have a solid foundation, and we must preserve that. We should then focus on creativity, inspiring students to develop an interest in mathematics, to make them feel that 'math is fun.' I hope that in Chinese classrooms, we can cultivate a large number of world-class mathematicians."

▸ Feel the Beauty of Mathematics

Chern often said that there aren't many wonderful things in the world, and mathematics is one of them. In 2003, the year of Guiwei (Water Sheep), followed by the year of Jiashen (Wood Monkey), he had the idea to design a calendar titled *The Beauty of Mathematics*. He personally conceived and designed it, aiming to present

3 See reference [1], p. 382.

the profound beauty of mathematics in an accessible way. The calendar featured twelve colorful pages, each dedicated to a different theme: complex numbers, platonic solids, Liu Hui and Zu Chongzhi, the calculation of pi, Gauss, conic sections, the double helix, the International Congress of Mathematicians, the development of computers, fractals, Maxwell's equations, and the Chinese Remainder Theorem. Each page included elegant, simple text that introduced key mathematical theorems and celebrated the great mathematicians of history, accompanied by intuitive graphics and photos that explained famous formulas and their applications. The calendar served as a brief yet illuminating introduction to mathematics and its history.

Chern had a particular affection for complex numbers, and he chose them as the theme for the first page of the calendar. His Chern classes was a topological invariant of the complex vector bundle. He once said: "Complex numbers are truly a magical field. For instance, with complex numbers, any algebraic equation can be solved, whereas in the real number domain, this is not possible … My focus is on 'complex' structures. 'Complex bundles' are simpler than 'real bundles.' In algebra, the complex number field is simple, and in group theory, the complex linear group shares similar simplicity. This simplicity is likely what gives complex vector bundles their significance."

He also remarked, "The role of complex numbers in geometry is a profound mystery to me. They are both incredibly beautiful and elegantly unified." He expressed regret that ancient Chinese mathematicians did not discover complex numbers, but he believed that his pioneering work in complex geometry could help make up for this historical gap.

China's Aspiration to Become a Mathematically Advanced Country in the 21st Century

In the spring of 1980, during a lecture at Peking University, Chern shared a profound insight: "Mathematics is an ancient and venerable discipline. In modern society, due to advancements in science and technology and the growing complexity of social systems, mathematics has become an essential pillar of education." He added, "Looking at the history of mathematics, we are now living in its golden age." "Why do we study math? The answer is simple. Because all other sciences rely on it." "Modern Chinese mathematics began later than that of Japan, but the scope and impact of the work of Chinese mathematicians are both broad and remarkable." In the end, Chern visioned: "We hope to see China emerge as a mathematically advanced country in the 21st century."

At the 21st Century China Mathematics Outlook Symposium, held at the NKIM in 1988, the idea of China becoming a "mathematically advanced country" was referred to as the "Shiing-Shen Chern Conjecture." So, what did it mean for China to be so? Chern explained: "The goal of Chinese mathematics is to achieve equality and independence. Our competition with Western mathematics is not about being the best, as in a race where only first and second place count. Rather, we must aim for genuine equality, where each side has its strengths and weaknesses. Furthermore, we seek independence—Chinese mathematics doesn't need to follow the same direction as Western mathematics, but it must be of the same caliber."

How, then, can China secure an advantage in mathematics in the 21st century? Chern said: "The solution is simple, to cultivate talent, to bring capable individuals into mathematics. We must identify and nurture exceptional young minds who will advance in the field. More concretely, we should establish ten world-class mathematical institutes in China." Guided by this vision, Chern founded the NKIM in Tianjin.

By the dawn of the 21st century, China's mathematics had made impressive strides. The 2002 International Congress of Mathematicians, held in Beijing, saw more than 20 mathematicians, educated in China, presenting 45-minute lectures. At this juncture, Chern believed that China had made a promising start toward realizing its goal of becoming a "mathematically advanced country." However, recognizing that the 21st century spans a full century, he emphasized that the next objective must be to establish China as a "world-leading mathematical nation." He thoughtfully remarked: "What we should think about now is how to raise China's level to match that of developed nations after the conference. This is far more important than simply presenting dozens of reports at the conference."

Indeed, while Chinese mathematics has made significant progress, it still lags behind the leading mathematical powers. The challenge now is to close these gaps and lead the development of mathematics in some areas, with the ultimate aim of establishing China as a true mathematically advanced country and world-leading mathematical nation. This should be the goal for future generations of Chinese mathematicians.

References

1. Zhang Dianzhou, Wang Shanping. The biography of Shiing-Shen Chern (revised edition). Tianjin: Nankai University Press; 2011.
2. Zhang Dianzhou, Wang Shanping. The collected works of Shiing-Shen Chern. Shanghai: East China Normal University Press; 2002.
3. Shing-tung Yau, Yang Le, Ji Lizhen, Wang Shanping. Mathematics and humanities (volume 3). Beijing: Higher Education Press; 2011.
4. Wu Wenjun, Ge Molin. Shiing-Shen Chern and Chinese mathematics. Tianjin: Nankai University Press; 2007.

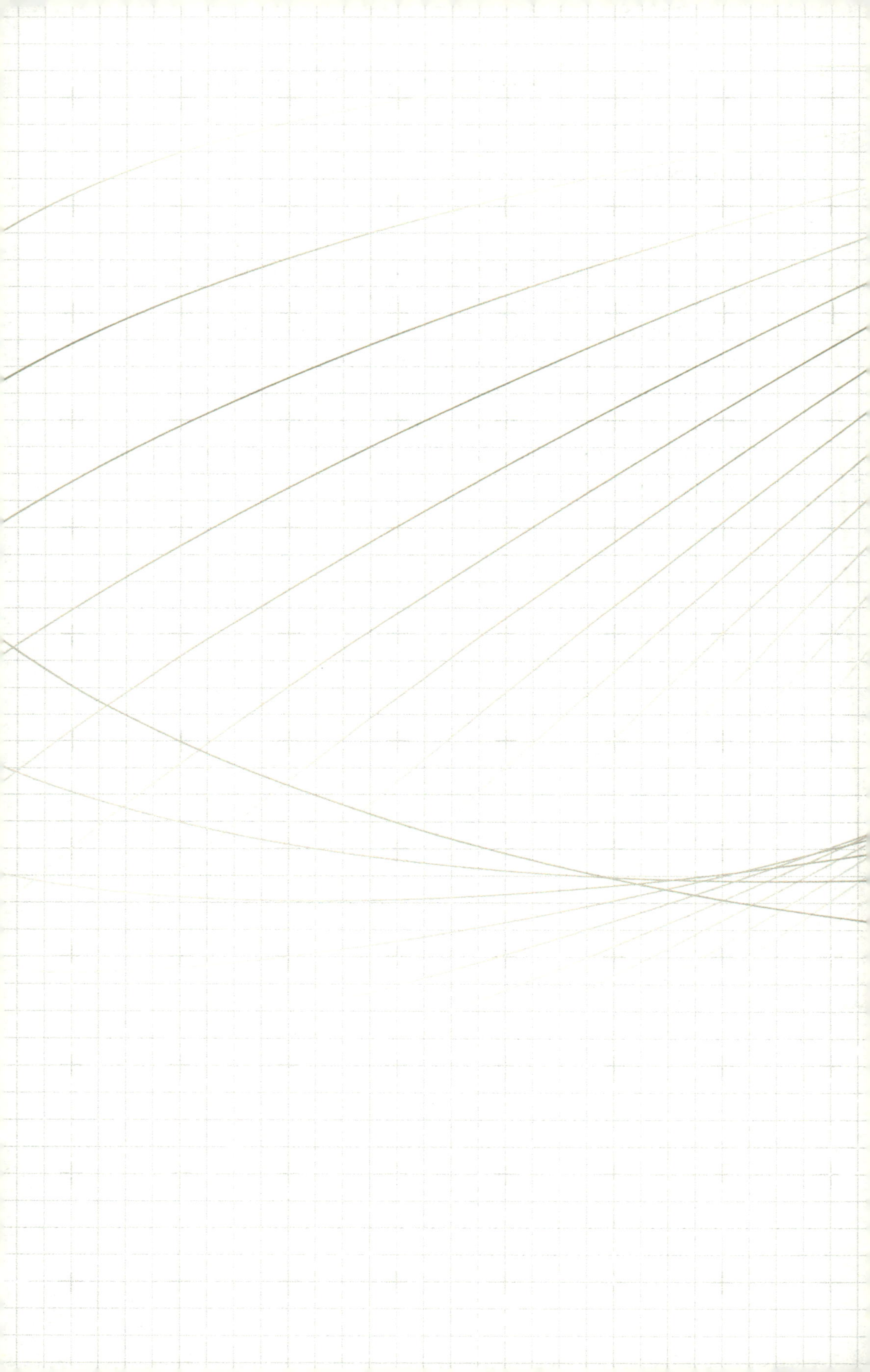

On Mathematics

Visions for the Future of Chinese Mathematics*

Mathematics is an ancient and venerable discipline. In modern society, due to advancements in science and technology and the growing complexity of social systems, mathematics has become an essential pillar of education. The widespread use of computers has also spurred the emergence of many new mathematical questions. Looking at the thousand-year history of mathematics, we are now living in its golden age. The number of active mathematicians today is unparalleled in history—there are now more living mathematicians than the sum of all mathematicians who have ever lived. In many countries, a growing number of people are supported by the state to pursue mathematical research, an unprecedented phenomenon. The result of this support has been tremendous progress in mathematics, evolving at a breathtaking pace. What is taught in the leading universities and research institutes today often did not exist just two or three decades ago.

Unlike music or the visual arts, the weakness of mathematics is beyond understandable to the public. While the efforts made by mathematicians to popularize the subject are commendable, the general public remains somewhat distanced from this field, which is not conducive to its growth. Mathematics, like a living organism, thrives only with constant, long-term progress. If progress stops, it perishes.

Why do we study math? The answer is simple. Because all other sciences rely on it. Let me share a story to illustrate this: Two classmates, A and B, went their separate ways after graduation. One day, they met again, and A asked B, "What have you been working on?" B replied, "I've been studying statistics, particularly population issues." A looked at B's paper and saw many formulas, especially the recurring symbol π. A said, "I remember learning about this symbol in school. It

---

* This article is a revised version of Mr. Chern's speech delivered in the spring of 1980 at Peking University, Nankai University, and Jinan University. It was originally published in *Chinese Journal of Nature* 4, no. 1 (1981).

represents the ratio of a circle's circumference to its diameter. I never imagined it would have anything to do with population studies!"

In China, the process of modernization is often compared to a second Long March. Mathematics is but a small part of this grand journey. The great French mathematician Poincaré once said, "In the battle of science, the enemy is always retreating." Therefore, this second Long March is far more fortunate than the first. However, the difficulties remain immense, as modern science is vast, and constantly evolving. Achieving success will require enduring effort.

In this march toward modernization, mathematics has certain advantages. First, it requires very little equipment. Second, its research directions are not highly concentrated, allowing smaller countries and institutions to create vibrant mathematical environments and foster respected mathematicians. Poland and Finland are excellent examples of this.

Mathematics is typically divided into pure and applied branches, but the line between them is not always clear. Often, pure mathematics has unexpected applications. For instance, the differential geometry required for Einstein's theory of general relativity was developed by Riemann over sixty years ago. The operator theory necessary for quantum mechanics was established long before by Hilbert. Recently, the quest for a unified field theory has been a hot topic in theoretical physics. Last year, Salam and Weinberg won the Nobel Prize for unifying the electromagnetic and weak forces. The mathematical foundation for this work was Yang-Mills gauge theory. In differential geometry, this theory is known as connection, and its geometric and topological properties have been a major focus of research for over three decades.

Differential geometry is the application of calculus to geometry, and I must highlight its impact on molecular biology. It is known that DNA's structure is a double helix, and the study of its total curvature uses J. White's formula, which has become a cornerstone of experimental molecular biology today.

These contributions, groundbreaking in pure mathematics, have become foundational tools in applied mathematics, representing the highest achievements in the field.

Modern Chinese mathematics began later than that of Japan, but the scope and impact of the work of Chinese mathematicians are both broad and remarkable. However, the number of mathematicians remains relatively small. By comparison, the American Mathematical Society has nearly ten thousand members!

To propel Chinese mathematics forward, I believe we should focus on two key points:

First, we must cultivate a young, ambitious generation of mathematicians. These individuals should be driven, confident, willing to sacrifice personal fame and gain, and dedicated to surpassing their predecessors. If Chinese mathematics achieves global leadership, the names of the mathematicians leading this movement will be those who are currently unknown.

Second, national support is critical. While mathematics does not require vast amounts of equipment, it does need sufficient material resources—such as access to libraries, ample research space, and greater domestic and international exchange. No one person can achieve everything alone; we must unite and work collectively toward this shared mission.

We hope to see China emerge as a mathematically advanced country in the 21st century.

Fifty Years of World Mathematics*

As we celebrate the 50th anniversary of the Chinese Mathematical Society today, I feel incredibly fortunate to be part of this grand occasion. It is truly an honor to be invited to deliver this speech. Just now, Mr. Su spoke with enthusiasm about the tremendous progress of Chinese mathematics over the past fifty years, and it was both exciting and inspiring to hear. The ultimate goal of Chinese mathematics is to achieve international excellence, and in light of this, I would like to take this opportunity to reflect on the developments in world mathematics over the last five decades.

Let me begin by sharing a bit of personal experience. In 1935, I studied in Hamburg, Germany, a city that was then a hub of mathematical research in the country. I recall that algebraic topology was taught by the eminent algebraist Emil Artin. At the time, the best book on the subject was a newly published work by Seifert and Threlfall (which was later translated into Chinese by Mr. Jiang Zehan). This book is now regarded as an introductory text for graduate courses, and its first half covers material that has become foundational knowledge for post-graduate students. In fact, our current understanding of topology probably exceeds the content of that book by tenfold. Many of the advanced topics covered in today's graduate programs were nonexistent fifty years ago.

Reflecting on fifty years of mathematics is no small feat—it feels like embarking on an immense journey, akin to recounting *The Twenty-Four Histories*. Thus, I will focus on the work of several prominent international mathematics awards and their recipients over this period. However, it is important to note that I will not attempt to cover all the groundbreaking contributions, as the field has expanded significantly

* In 1985, the Chinese Mathematical Society held the 50th Anniversary Conference of the Chinese Mathematical Society at Fudan University in Shanghai. This is Mr. Chern's speech at the opening ceremony. It was originally published in *Science* 38, no. 1 (1986).

over these years. The number of mathematicians has grown dramatically, and the speed at which mathematical ideas and content have evolved is unprecedented.

Today, the most prestigious international mathematical award is the Fields Medal, presented at the International Congress of Mathematicians. Fields was a professor at the University of Toronto in Canada, and in his will, he established this prize to reward young mathematicians. The recipients of this award must be under 40 years old, and it is given once every four years. Since its inception in 1936, the Fields Medal has been awarded to 27 mathematicians. Among these, Chinese American mathematician Shing-tung Yau received the medal in 1982–1983.

In 1979, the Wolf Prize was established by a fund in Israel, offering substantial monetary rewards in six categories, including mathematics. Unlike the Fields Medal, there is no age limit for this prize, and it is awarded based on a mathematician's entire body of work. The Wolf Prize is awarded annually, and to date, 14 mathematicians have received the prize, all of whom are over 60 years old. I had the honor of being one of the recipients in 1983–1984.

Two individuals have received both the Fields Medal and the Wolf Prize: Lars Valerian Ahlfors, originally from Finland and now a US citizen, and Kunihiko Kodaira from Japan.

What are the fields of work of these distinguished prize winners? Their areas of expertise are diverse and difficult to categorize into any one specific branch. After reviewing their achievements, I found that most of their work relates to topology, analysis, or number theory. In general, they have followed the great traditions of mathematics from Gauss, blending pure and applied mathematics into a unified whole. We could call this core mathematics. There is a continuity to mathematics—spanning from Euclid to Gauss and extending through these 41 award winners and their friends, sharing a common intellectual spirit.

However, the scope of mathematics has expanded greatly. In the 18th and 19th centuries, the fundamental mathematical problems were concerned with understanding infinity: both infinitesimal and infinite. Thus, analysis was the heart of mathematics, focused on a single variable. One of the greatest mathematical breakthroughs was the introduction of complex numbers, where a single complex number represents two real numbers. This led to the study of complex functions, which involve two-dimensional and higher-dimensional spaces, and consequently, the development of topology became inevitable.

Professor Ahlfors, for instance, worked on complex analysis, while Professor Kodaira focused on transcendental algebraic geometry. Both of these areas are rooted in complex numbers, which play a pivotal role in modern mathematics.

The mathematics of the 19th century was one-dimensional, but the mathematics of the 20th century has become high-dimensional, involving multiple spaces. As a result, topology has become foundational, providing a unified perspective that can be applied to many other fields of mathematics.

Number theory is the most profound branch of applied mathematics. While elementary number theory is beautiful, algebraic number theory is truly deep. It is intricately linked with algebraic geometry. All of this constitutes the highest level of core mathematics.

I must also mention the awards in applied mathematics. A notable recent prize is the Kyoto Prize from Japan, which offers a significant monetary reward. The recipients include two pioneers in information theory, Claude Shannon, and in systems theory, Rudolf Kalman. With advancements in science and technology, and the rise of computers, many new areas in applied mathematics have emerged. On the other hand, the traditional relationship between mathematics and physics remains extraordinarily close. Examples include the connection between gauge theory and fiber bundles, as well as string theory and the Kac-Moody algebras. Another striking example is the discovery of solitary wave solutions to evolution equations, first observed through computer simulations.

Both core mathematics and applied mathematics continue to evolve and expand, seamlessly blending into a unified field. Modern mathematics is filled with vitality and potential.

Dear colleagues, as we celebrate the achievements of the past fifty years, we must also look forward and plan for the next fifty. In recent years, I have seen many young and promising mathematicians, and I am convinced that we are witnessing the beginning of a golden age for Chinese mathematics. I firmly believe that within the next fifty years (or even sooner), Chinese mathematicians will win the highest international honors based on their work conducted within China. Chinese mathematics is an inseparable part of world mathematics.

To conclude, I would like to end with a few words from Mr. Su: "We must uphold rigorous scholarship and united effort. We must promote democracy, encourage broad participation, respect one another, collaborate, and work together to build a stronger community for the future of mathematics."

Young Mathematicians Should Immerse Themselves in the Works of Great Masters*

In an interview[1] with *The Mathematical Intelligencer* (vol. 6, 1984, pp. 9–19), Atiyah was asked which mathematician he most admired. His response was clear: "This is an easy question to answer. The mathematician I most admire is H. Weyl. He was interested in group theory, representation theory, differential equations, the spectral properties of differential equations, differential geometry, and theoretical physics. Almost everything I have done, in a conceptual sense, has its roots in his work. I also fully embrace his mathematical philosophy and share his views on what makes something truly fascinating in mathematics." We have found out that this profound mathematical philosophy and spirit have been preserved and carried forward in the collected works that follow.

I encourage my Chinese colleagues and students to treat this collection as an advanced "textbook." No matter how refined or complete new interpretations of a subject may be, the original papers often offer a more direct and incisive insight. When I was younger, I heeded the advice of my mentors to read the works of great mathematicians such as H. Poincaré, D. Hilbert, F. Klein, and A. Hurwitz, and those

---

* This is the preface written by Mr. Chern for the mainland edition of *Michael Atiyah Collected Works* (Oxford: Oxford Press, 1988), translated by Zhang Weiping (see Zhang Weiping, "The Emergence of Index Theorem in China: In Memory of Shiing-Shen Chern," *Advanced Mathematics Research* 14, no. 5 [2011]: 1–3). The editors of this book have made slight revisions to the translation and added a new title.

1 It has been translated into Chinese by Wang Qiming as "Atiyah's Visit" and included in *The Unity of Mathematics* (Yuan Xiangdong, ed. [Dalian University of Technology Press, 2014]).

works deeply enriched my understanding. Later, I immersed myself further in the writings of W. Blaschke, Elie Cartan, and H. Hopf.

This approach also echoes the rich traditions of Chinese scholarship: ancient scholars were expected to study the *Analects of Confucius*, the prose of Han Yu, and the poetry of Du Fu. I earnestly hope that these collected papers are not simply placed on a shelf as decorations, but that they are read and reread, well-worn from frequent study, by young mathematicians.

The Mathematical Capability of the Chinese Nation: No Longer in Need of Proof*

First, I would like to thank Mr. Wu Wenjun for his outstanding presentation. In just one hour, we learned so much about topology—its history, its fundamental concepts—and we also gained insight into the significant contributions of Professors Jiang Boju and Li Banghe in the field. I'd like to add a small note regarding Mr. Wu's mention of the "plum blossom knot," a knot in space that cannot be untied. He drew the diagram. Though this appears to be a simple geometric problem, it has profound implications in physics. This concept plays a crucial role in statistical mechanics, particularly in Chen-Ning Yang's equations.

Allow me to further mention that Professor Yang, established a theoretical physics group at the NKIM, and there along with Mr. Ge Molin, compiled a book on statistical mechanics. The final article in that book is by Witten, who is regarded as one of the most extraordinary theoretical physicists and mathematicians of our time. His article talks about establishing a unified field theory that encompasses all types of fields, and its foundation rests on topology and differential geometry. Although Witten is a physicist, his article is very difficult for mathematicians, offering a series of statements that are casually made without rigorous mathematical proof. Now, the most prominent geometers and topologists worldwide are tirelessly working through Witten's article, attempting to derive proofs from his words, and often finding that their reasoning is correct. However, the proofs remain elusive and, at least for now, are incomplete. Personally, I have devoted much of my time to grappling with his article, much like the way mathematicians of the past, such as

---

* This is Mr. Chern's speech at the second "Shiing-Shen Chern Prize" award ceremony, with the title added by the editors of this book. Originally published in the *Communications of the Chinese Mathematical Society* in November 1989.

Poincaré, dedicated themselves to understanding and proving his topological ideas. As Mr. Wu mentioned, the approach back then was to painstakingly study Poincaré's papers—what appeared to be simple ideas required immense time to grasp and even longer to prove. Witten's article is the concluding piece in the book on statistical mechanics, edited by Chen-Ning Yang and Ge Molin.

Now, adding another topic, I would like to touch upon fixed-point theory, which is also part of mathematical history. A result in this field is the so-called Fundamental Theorem of Algebra, which states that every algebraic equation has a complex solution. Many may not realize that this was once a challenging problem. Gauss provided several proofs for it, but before Gauss, the great mathematician Euler could not prove it. Euler was only able to prove special cases, and it wasn't until Gauss introduced the concept of topology that he was able to prove it in its entirety.

Let me now speak briefly about the process behind the establishment of this award. One day, while I was at my home in California, I received a call from Mr. Chen-Ning Yang, who told me that Mr. Jimmy Liu from Hong Kong wished to speak with me. After our conversation, Mr. Liu proposed the idea of creating this award, which I consider an immense honor. It is also a testament to Mr. Liu's dedication to advancing the mathematics of our homeland, and I deeply respect his contribution. Naturally, I agreed to the proposal. It came as a bit of a surprise, as no prize has ever been named after me before. This marks the second time this award is being given, and as we've learned from Mr. Wu's report, the selection of this year's awardees has been exceptionally well made. The two winners this year (Li Banghe and Jiang Boju), along with the two from last time (with the unfortunate passing of Professor Zhong Jiaqing), have all reached international standards. They would be fully deserving of similar awards on the global stage. Therefore, Chinese mathematics no longer needs to suffer from any sense of inferiority; we have reached a level that matches the best in the world, though there is always room for improvement.

I'd also like to mention an important piece of news this year: China won first place in the International Mathematical Olympiad. Two years ago, China placed fourth, and last year, we achieved second place. This year, we proudly took first place. This is an incredible accomplishment, especially considering that around 40 to 50 countries participated. I understand that one of the problems in the competition was particularly difficult—at least I couldn't solve it myself. However, young Chinese students were able to solve it and secure this victory, outperforming other strong countries like the Soviet Union, Romania, East and West Germany, and the United States. This is a moment of great pride for us all. I believe we must continue to nurture these young talents, ensuring that some of them go on to become world-class mathematicians. Perhaps in ten years, Chinese mathematicians will win the Fields Medal for work done on Chinese soil. While Professor Shing-tung Yau won the Fields Medal, his work was carried out in the United States. This is a thrilling prospect.

Allow me to share another story. This summer, the American Mathematical Society held a several complex variables symposium in California, spanning three weeks. If you attend it, you will quickly realize that Chinese mathematicians, especially Chinese descendants, are leading the way. Mathematicians like Xiao Yintang, Shing-tung Yau, Mo Yiming, and Tian Gang are all at the forefront of the discussions. Their work receives the full attention of the symposium and constitutes some of the most important contributions to the field. Beyond these individuals, many young Chinese mathematicians, their students, are also making significant strides. So, if you attend it, it almost feels as if you are attending a meeting in China, as the majority of participants are Chinese. This is a clear testament to China's leading position in the field of several complex variables. This achievement is no small matter.

In recent years, with the support of the Ministry of Education, we have continued to send outstanding students abroad, particularly to the United States. Many of them have excelled and found themselves in top-tier institutions such as Princeton and Harvard, where Chinese students are now counted the best in mathematics departments. Even in mid-tier schools, where it's difficult to attract domestic talent, Chinese students shine. Many of these students, when they present their work, find themselves surrounded by their peers—other Chinese students.

I often say that we should aim to transform China into a mathematically advanced country. If we consider all ethnic Chinese worldwide as part of the broader Chinese community, then in a world unified under China's banner, we would already be a mathematical giant. Indeed, we have already achieved considerable success. As Comrade Wang Yuan mentioned earlier, in our youth—especially mine—there was a sense of inferiority. He recalled seeing my work in Steenrod's book. I vividly remember that in a book by Fubini, the only Chinese paper cited was that of my teacher, Sun Tang. It wasn't an extraordinary paper—just a brief mention in the literature. At that time, there were no other Chinese names in the references. Naturally, we felt a certain inferiority. But now, I believe that sense of self-doubt is no longer necessary. The mathematical capability of the Chinese nation is undeniable—it is now evident for all to see. The challenge ahead is to continue advancing.

I have a few concrete suggestions to share. Earlier, I mentioned the success of China's team in the International Mathematical Olympiad, where many young Chinese students have demonstrated extraordinary mathematical abilities. Mr. Qiu Zonghu, with whom I spoke, pointed out that China is not only capable of sending one competitive team but could easily field another equally talented one. However, I want to stress that while these competitions are important, they should not be seen as the ultimate measure of mathematical achievement. Historically, Oxford and Cambridge held prestigious mathematics competitions, which included very difficult problems. According to statistics, few of the greatest mathematicians were the winners of these competitions. The only exception was Maxwell.

From what I understand, people around the world are increasingly reluctant to study mathematics. Even in the Soviet Union, students are shying away from the subject. But at this moment, when others are stepping away, it is precisely the right time to study mathematics. As Mr. Jimmy Liu would surely agree, mathematics will soon become an invaluable skill. We must continue to nurture capable young individuals who will dedicate themselves to mathematics. This will not only bring honor to China but will also be of immense significance for their future careers. We do not want everyone to study mathematics—too many mathematicians might not be ideal—but we do need a substantial number of capable individuals in the field.

A second point I want to make is that Chinese mathematics must become independent. Now that we have gained confidence, we no longer need to follow the paths laid out by other countries or their great masters. To stand on our own, it is essential that we develop a deep understanding of mathematics itself. We need to ask: Why study mathematics? What are the current challenges in the field? What developments can we expect in the next ten, twenty, or even fifty years? Ideas are not fixed; what seems true today may change tomorrow. But to allow for such change, we must first have a holistic view of mathematics. One area I believe China can develop further is in the history of mathematics. This is a field that Western countries have not given enough attention to, and I believe this presents a unique opportunity for us. Every discipline develops according to its historical trajectory, and understanding that history is crucial to understanding the subject itself.

I know that many Chinese scholars are already deeply interested in the history of mathematics, and I believe we should encourage their work. Writing the history of mathematics is no small task—it requires significant effort, particularly when dealing with earlier periods. Researching ancient or medieval mathematics, or even 19th-century mathematics, is difficult and requires years of training in languages like Latin. But I believe the history of modern mathematics is more accessible, and I would encourage scholars to pursue this area. I remember an idea proposed after China's reforms: Many top scientists, including distinguished mathematicians, have visited China. Why not take this opportunity to write about these visits? Several of us—especially those interested in the history of mathematics—could write about these interactions. Fifty years from now, these accounts will be invaluable historical resources.

As for the other developments in mathematics, much of the current work is focused on discrete mathematics. In contrast, we older generations were trained in topology. Mastering topology is no small task—it demands years of dedicated study to truly understand, and the same applies to number theory, where it may take a decade of hard work just to enter the field. Discrete mathematics, however, presents a wider array of problems, particularly with the recent surge in computing advancements, which have given rise to a host of exciting new challenges in the field.

In conclusion, my message is clear: We should no longer merely follow in the footsteps of others. Instead, we must find new, meaningful directions for our work. Chinese mathematicians should take the lead in these areas, carving out a path of our own, rather than simply adhering to the established trends.

In addition to theoretical concerns, there are practical matters that I believe we can address within China's current context. One such initiative is the organization of more regional mathematics conferences. Given the vast size of China, hosting a national conference demands considerable financial resources. In contrast, regional conferences—where we invite a few experts from different areas to give talks—would be far more cost-effective. Last year, we organized a 21st-century mathematics conference, supported by a modest grant from Comrade Li Tieying. He even mentioned that we could hope for greater funding in the future.

Why emphasize regional conferences? Mr. Wu Wenjun's report serves as an exemplary model. He demonstrated how, within a limited amount of time, one can give a concise yet thorough overview of an important subject. It's crucial that, across China, those interested in mathematics attain a certain level of expertise. Earlier, I spoke about exploring new areas and taking different approaches in mathematics. But maintaining a high level of understanding is equally important. Without this, even those who are simply curious about mathematics might spend years on unsolved problems like Fermat's Last Theorem or the Goldbach Conjecture, without truly understanding the nature of mathematical proof. What we need to continuously strive for is the cultivation of essential mathematical skills—basic training and a fundamental understanding of what constitutes mathematical proof. This is a critical step that everyone should be working on.

Last, looking further ahead, I've been reflecting on the recent decline in international exchange opportunities. This brings to mind Chairman Mao's concept of "self-reliance." In the end, we must rely on ourselves, especially the younger generation. Unfortunately, there are fewer young people fully dedicated to this cause, but I urge you all to pass this message along. Younger mathematicians should unite and pursue their own work—not relying excessively on foreign experts, nor even too much on domestic mentors. A striking example of this spirit is the Bourbaki school in France, which grew into a major intellectual school. This group of young, brilliant mathematicians was motivated by the fact that most of the prominent figures in Parisian mathematics, especially the professors at the University of Paris, were deeply focused on function theory. These established scholars were largely unaware of the emerging directions in mathematics. In response, the younger generation of French mathematicians, full of ambition and vision, decided to form their own school of thought and chart their own course, shaping a new era in mathematical development.

I can fully appreciate how difficult it is for scientists to make progress when distracted by superficial fame or reputation. The more one's name grows, the more one's attention may drift away from the essential work, leading to stagnation or regression. This is why it is critical for young mathematicians to be self-reliant. They must take charge of their own projects, seek out problems to solve, and engage in independent intellectual discussions.

I recall reading a book by the renowned French novelist Romain Rolland, which narrates the story of nameless German musicians. When a group of young German musicians traveled to Rome to study, the Romans dismissed them as barbarians, unworthy of understanding music. Yet, within a few short years, Germany gave birth to a remarkable generation of composers—Wagner, Beethoven, Bach, and many others. This story stands as a testament to the transformative power of dedication and ambition. With determination, especially in the realm of mathematics, even the greatest challenges can be overcome. As for material obstacles, we are fortunate to have the unwavering support of Mr. Jimmy Liu.

We must recognize that mathematics holds tremendous promise. Over the past two thousand years, it has become one of humanity's greatest disciplines. Today, mathematics is not only an intellectual pursuit but also underpins numerous fields, such as computer science, biology, and engineering. This is truly a remarkable phenomenon. Take, for example, the simple equation $x^2 + 1 = 0$. As a result, discussions like these, seemingly abstract, can actually have practical applications—something almost unimaginable. For example, in order for this equation to be meaningful, it requires complex numbers, and in turn, complex numbers do find applications. This is truly a fascinating aspect of mathematics. Therefore, the future of mathematics is incredibly promising. In the context of China, there is so much to be done.

With that, I conclude my speech for today.

Mathematics in the 21st Century*

I am deeply honored to have the opportunity to speak with you today. Allow me to begin with two stories.

We are all familiar with Euclid's *Elements*, a foundational work in mathematics completed over two millennia ago. It is an extraordinary contribution to humanity. The book covers much more than just geometry, incorporating elements of analysis and algebra. Its primary aim was to use reasoning to derive conclusions about geometry. In particular, chapter 13 deals with the number of surfaces on regular polyhedra. A regular polyhedron is a polyhedron where its surfaces coincide with each other, all surfaces meet at a single vertex, and the number of edges per surface is the same. In the case of regular polygons in a plane, they include shapes like equilateral triangles, squares, and so on. However, when extending the concept to three-dimensional space, the situation becomes more complex. There are far fewer regular polyhedra in space, with only five types: the tetrahedron, hexahedron (cube), octahedron, dodecahedron, and the largest one, the icosahedron. A friend of mine wrote a book that includes these beautiful geometric shapes, and I have a colored copy of it here.

Some might wonder, "What do mathematicians spend their time on? What's the point of studying these polyhedra?" But as I mentioned to Mr. Zhang Cunhao, titanium compounds in chemistry are related to the concept of regular polyhedra. This is a fascinating fact—after two thousand years, regular polyhedra have found applications in chemistry. In fact, some mathematicians are now studying the connection between polyhedra and molecular structures. It is well-known that viruses

---

* This article is a speech delivered by Mr. Chern on May 31, 1992, at the Academic Report Meeting in Commemoration of the 10th Anniversary of the National Natural Science Foundation of China. It was originally published in *Communications of the Chinese Mathematical Society* in June 1992.

in biology often take on polyhedral shapes. This demonstrates how what was once considered a "fantasy" by mathematicians has, over time, become highly practical.

Now, let me share another story that is often recounted. Two friends from high school, who had not seen each other in years, happened to run into one another one day. A asked B, "What have you been doing all these years?" B replied, "Studying population issues." Naturally, A was curious to see his friend's work, so he read B's paper on demography. To his surprise, the paper contained numerous appearances of π. He thought this was strange—after all, π is the ratio of the circumference of a circle to its diameter. How could it possibly have anything to do with population studies? This situation, much like the one with regular polyhedra, illustrates a crucial point: fundamental sciences, especially pure mathematics, are often difficult to predict in terms of their important future applications. If we demand that fundamental science be immediately applicable, we are displaying a shortsighted view.

Mathematicians often ponder problems at home, wondering why their ideas might prove useful. I believe the reason is clear—mathematics has a remarkably simple yet profoundly solid foundation. Its conclusions are derived from logical reasoning, which makes them entirely reliable. Logical reasoning provides a higher degree of certainty than empirical experiments. Because of its logical rigor, mathematics is a field built on a solid foundation, and this is one of the reasons it proves to be so useful.

There is another question: Why do so many diverse fields of study rely on the same mathematics? This remains a complex issue to understand. One possible explanation is that truly good mathematics is rare. There are only a few grand mountains in the world, and beautiful places are far from abundant. When you visit Beijing, for instance, the most famous landmarks are the Great Wall, the Temple of Heaven, and the Forbidden City—there aren't many options. Similarly, when it comes to applied mathematics, we often rely on just a handful of particularly brilliant mathematical concepts, and such good mathematics is not plentiful.

Today, I speak on the topic of mathematics in the 21st century—specifically, how Chinese mathematics can develop and how China can position itself to have significant advantages in certain areas of mathematics in the 21st century. The solution to this challenge is simple: We must cultivate talent, identify capable individuals, and provide opportunities for young people to develop in mathematics. To elaborate further, the goal is to establish ten world-class mathematical research institutes across China. Given the vastness of China, these institutes should not be limited to Beijing but should also be spread throughout other regions. In total, we might aim for about ten such institutes.

As for what qualifies as "world-class," this is not strictly defined. I can only say that, currently, the mathematics department at Nankai University does not yet meet this standard, and considerable effort is required to bring it up to global levels.

The foundation of Chinese science must be deeply rooted within China itself. Only after establishing strong roots here can science and technology begin to flourish. However, attracting talented individuals to work in mathematics is no simple task. In 1984, I began the process of establishing the NKIM. Initially, I thought the solution would be straightforward—simply invite talented people to work at the institute. However, due to various challenges, this proved to be more difficult than expected. To create a top-tier research institute, we need top-tier mathematicians. With top-tier mathematicians, even if the building is modest, the equipment is basic, and the resources are scarce, the institute can still be world-class. However, if the building is grand, the library is stocked with books, and expensive equipment is available, but there are no mathematicians to conduct high-level research, then none of it matters.

Recognizing this, I shifted my approach. I focused on training our own young talent, nurturing homegrown mathematicians, and sending them abroad for study. We also invited the best mathematicians from around the world to provide guidance and mentorship.

I am pleased to report that these efforts are beginning to bear fruit. For example, He Zhengxu went to the University of California, San Diego, to study under Friedman, a Fields Medalist and a prominent young leader in the field. Friedman himself has told me that He Zhengxu is his best student. He Zhengxu is now at Princeton. Another example is Wang Shuguang, who received a scholarship from the Wang Kuancheng Foundation and placed first in the selection exam. I introduced him to Oxford University, where he worked with Donaldson, one of the UK's leading mathematicians. Donaldson, who is now a professor at Oxford despite being under 30, has been instrumental in Wang Shuguang's development. Wang completed his doctoral thesis just a year ago. Then there is Wang Rongguang (not related), who also received a scholarship from the Wang Kuancheng Foundation. He went to Harvard University to study under Taubes, and this year, he finished his dissertation. Another example is Zhang Weiping, who studied under Bismut, one of France's most renowned young mathematicians (the other one is Connes). Zhang completed his PhD in just two years in Paris and is now a postdoctoral researcher at the Institute for Advanced Studies in Paris. I could mention many more, but time is limited.

Of the four mentioned above, Zhang Weiping has agreed to return to China next year and join us at Nankai. When Zhang returns, I hope that the government can offer support to retain such a talent. The decision for overseas students to return largely depends on the domestic environment—specifically, the treatment of accomplished scientists. Today, I won't delve into this issue, but I do want to emphasize that talent should be able to flow freely across borders. Those returning from Europe should have the option to work in the US if they choose, as current policies are relatively flexible. The ease of going abroad means that we must find ways to retain talent through appropriate policies. However, dealing with such policy issues is beyond my purview, as my focus remains on mathematics.

Next, let's talk about mainstream and non-mainstream mathematics. As we know, mathematics is a highly versatile discipline. For example, it doesn't require much in the way of equipment. Thanks to email and the ease of accessing information, mathematics remains a highly personal endeavor, unlike other disciplines that heavily rely on equipment. In other fields, people compete to work in the most important directions, making their contributions in mainstream areas. Mathematics, however, is different. Since there are many directions in mathematics, and it is a personal endeavor, not everyone needs to focus on mainstream mathematics. In fact, I believe we should encourage people to explore non-mainstream areas as well. Often, what is not considered mainstream today may become central in a few years.

Allow me to share a personal experience. In 1943, I was teaching at the National Southwest Associated University, and Chen-Ning Yang was a graduate student there. That year, I was invited to the Institute for Advanced Study of Princeton University, where Yang later became a professor. Near Princeton is a small town called New Brunswick, which houses Rutgers University. In August, I traveled to Princeton, and shortly after, I attended the American Mathematical Society's Summer Meeting in New Brunswick. Since it was nearby, I attended the talks and caught up with friends. On one occasion, I was chatting with a highly respected American mathematician who asked what I was working on. I replied, differential geometry. He immediately said, "It is dead." This was in 1943, but by the end of the war, differential geometry had become one of the mainstream areas of mathematics.

Therefore, I believe that mathematicians have the potential to discover directions that, while not mainstream today, are highly meaningful and have great promise for the future. Mainstream fields are already dominated by many of the world's leading experts, and large amounts of funding are poured into them. Competing directly with such forces may not be realistic. Instead, it may be more productive to focus on other areas that are equally significant. My hope is that Chinese mathematics can establish itself in certain fields, excel, and develop its own distinctive characteristics. This approach is not without precedent in history. For instance, before World War II, Poland focused on logic and point-set topology. Based on simple axioms, they derived numerous significant results. Finland, too, achieved notable success in complex analysis, and continues to lead the world in the extension of quasi-conformal mappings. Their success stemmed from choosing to work in areas where other countries were not involved, ultimately positioning themselves at the forefront of those fields. There are many such examples.

Earlier, I mentioned the goal of establishing ten world-class research institutes. What would it take for these institutes to truly attain a world-class standard?

First, we must offer advanced foundational courses. When students arrive, they need foundational training, which necessitates offering courses of the highest caliber. This "foundational training" comprises two components. The first is the development of logical reasoning skills. A student must learn to distinguish between correct and incorrect reasoning, ensuring that every step in a proof is sound. Rushing to conclusions with sloppy work leads only to errors. The second component involves a broad understanding of mathematics. A student should be able to form a comprehensive judgment of the discipline as a whole. In the past, analysis was central to mathematics. Since the 20th century, algebra, including group theory and representation theory, became more prominent, followed by topology and other areas. A strong research center should be able to offer these basic courses regularly, at least once every two years. I believe that this is within China's reach. It comes down to two things: first, offering key courses at research institutes and providing scholarships; second, inviting foreign experts, even retired ones, to teach. Bringing in retired mathematicians from abroad costs little, as they already receive pensions. By providing for their living conditions and arranging a bit of tourism, they can contribute significantly. This way, an institute can offer a comprehensive and enriching curriculum.

Second, we must recruit exceptional students. Every year, we send top students to participate in the International Mathematical Olympiad, and many of them excel. While excelling in high school mathematics does not necessarily translate into a future in mathematical research, I hope that some of these students will pursue mathematics and achieve great things. Just yesterday, I met with some of the award-winning students from Beijing's math competitions. I told them that pursuing mathematics would offer them a promising future. In ten or twenty years, the world will certainly face a shortage of mathematical talent. Today's young generation is less inclined to study mathematics, which will inevitably result in a talent shortage. I can understand why students may shy away from mathematics—it is difficult, and there is no guarantee of success. After years of hard study, many students still find themselves far from becoming mathematicians. Meanwhile, other fields like computer science are competing for talent. A successful career in software engineering demands great skill, but the preparatory knowledge required is much less than that needed for mathematics. Mathematicians often feel like they are constantly learning, with an endless stream of books to read. As a result, many gifted individuals are drawn to the tech industry. Some math PhDs, after graduating, choose to work in the stock market, writing algorithms to predict stock market movements so that wise investment decisions are made. While they may still be employees, their salaries often far exceed those of university professors. Thus, the loss of mathematical talent is a global issue.

In contrast, China's situation is more optimistic because, even if some talent is lost, we can still cultivate more. Those who are lost and achieve financial success often return to help build the infrastructure for mathematics. Ultimately, we must adopt a particular mindset: China should become a factory for producing mathematicians. Those who leave should be encouraged to return, and if they don't, we should continue to send them abroad. China has an abundance of talent, and sending some abroad to make a global impact is worthwhile. What we must focus on is investing modestly to build a strong foundation and offer quality courses. Once the foundation is solid, whether those who leave return will become less important. This is my preliminary vision for the future.

For example, participants in the International Mathematical Olympiad are typically outstanding in mathematics. If they enter a university mathematics department, I recommend offering them scholarships right away. Though the financial reward may be modest, the impact can be significant, and it serves as an encouragement for others. Chinese students tend to listen to their parents and teachers, and if a child shows talent in mathematics, a little persuasion from their family and mentors can often set them on the path to a mathematics career.

For gifted students in mathematics, studying abroad is just a matter of time. If they excel in mathematics domestically, going abroad becomes relatively easy. If they perform well at home, they can skip graduate school and even become professors abroad right away. China already has the conditions to produce world-class mathematicians, and we must have confidence in this.

In terms of student development, I advocate for mobility. In the 19th century, Germany was, undoubtedly, the world leader in mathematics. German university students had the freedom to register at any university. One semester, they might attend lectures by Weierstrass in Berlin, and the next, they could attend Schwarz's class in Göttingen, with complete academic freedom. Professors also enjoyed this mobility. For example, Berlin University had Planck and Einstein, and if a theoretical physicist found no prospects at Berlin, they could easily move to another institution to start fresh. I hope that Chinese students and professors will have similar opportunities for mobility. Professors could teach at different universities for half of the year, and faculty at various mathematics institutes could exchange positions as well.

Now, let me discuss a bit more about mathematics. As I mentioned earlier, choosing a research direction in mathematics does not have to align with the mainstream. You can pursue branches that particularly interest you. However, A mathematician should understand what constitutes good mathematics and what is bad or less refined mathematics. Some mathematics is groundbreaking and has promising development potential—that is good mathematics. There is also mathematics that is quite interesting but gradually turns into a mere game. Choosing a good research direction is crucial.

Let me illustrate this with an example. Have you heard of Napoleon's Theorem? It might not have any real connection to Napoleon, but it's quite an intriguing result. The theorem states that for any triangle, if equilateral triangles are constructed on its sides, the centroids of these triangles will form an equilateral triangle. If the equilateral triangles are constructed inward instead, two solutions emerge. This type of mathematics is not considered "good" because it offers little opportunity for further development. Of course, if you're looking to unwind and want to ponder on this question casually, this could be enjoyable—almost like a game.

So, what does "good" mathematics look like? Take solving equations, for example. Solving equations is the cornerstone of mathematical inquiry.

Solving linear equations is easy, but quadratic equations are different. For example, the equation $x^2 - 1 = 0$ has real solutions, while $x^2 + 1 = 0$ does not. Eventually, the introduction of complex numbers allows us to explore the complex solutions to equations. The famous Fundamental Theorem of Algebra asserts that every polynomial equation of degree n has a solution in the complex number system. This issue has a long and storied history. The great mathematician Euler pondered this problem. Though Euler did not have a professor position and faced financial difficulties, the German emperor, adamant about having the world's best mathematician, invited him to work at the royal court. Despite his efforts, Euler was unable to prove the Fundamental Theorem of Algebra. It was Gauss, however, who eventually discovered the connection between complex numbers and topology, leading to a new understanding of the subject. A complex number with a modulus equal to 1 corresponds to a unit circle, and on this circle, a wealth of intricate patterns can unfold. Gauss was the first to prove the Fundamental Theorem of Algebra, offering not just one, but multiple proofs.

When we move from solving $f(x) = 0$ to $f(x,y) = 0$, we enter the study of curves, which may not have solutions—there may not even be a zero point at all. And that's when the variation unfolds. If we treat both x and y as complex numbers in equation $f(x,y) = 0$, the two complex numbers correspond to a four-dimensional space, leading us into the intricate world of Riemann surfaces in complex analysis. Solving the original equation $f(x,y) = 0$ through the lens of Riemann surfaces requires a wealth of mathematical knowledge, including the crucial concept of (genus) g. When we interpret the solutions to $f(x,y) = 0$ as a curved surface, the number of circles on the curved surface is linked to the (genus) g, which distinguishes different variation of the curved, spherical, and torus surfaces.

Furthermore, we can explore even more fascinating variations. For example, if we assume the coefficients of $f(x,y) = 0$ to be integers, we can look for its integer solutions. This adds another layer of complexity: When $g = 0$, there are infinitely many integer solutions. When $g = 1$, certain special properties arise. When $g > 1$, a groundbreaking result by German mathematician Faltings in 1984–1985 proved that the equation $f(x,y) = 0$ has at most a finite number of integer solutions. This result is directly tied to Fermat's Last Theorem, which states that the equation $x^n + y^n = z^n$ $(n \geq 3)$ has no positive integer solutions. Although this doesn't fully resolve the Fermat problem, it represents a monumental leap forward.

Mathematics, indeed, has the power to uncover deep insights. I consider number theory to be a branch of applied mathematics—applying mathematical techniques to explore the properties of integers. In my view, there are two central pillars of mathematics: number theory and theoretical physics. There is also a wide application of mathematics in theoretical physics.

In this brief hour, I cannot cover all the mathematical intricacies. However, there is one final topic I wish to touch upon: the most exciting developments in mathematics today—the current mainstream mathematics. As I mentioned earlier, I am not an advocate for everyone rushing to focus solely on mainstream mathematics, but it certainly holds great importance. Mainstream mathematics is characterized by monumental contributions—deep theorems with broad and far-reaching implications, whose proofs require extraordinary effort. If I had to select a contemporary example of such a contribution, it would undoubtedly be the Atiyah-Singer Index Theorem. Atiyah, the president of the Royal Society, visited Beijing last month and delivered a lecture on this subject. The index theorem can be seen as the modern extension of many of the ideas I have discussed, taking algebraic equations, Riemann surfaces, and genus theory, and expanding them from lower dimensions to higher and even infinite dimensions.

Thus, I believe that mathematical research is not just profound, difficult, and powerful, but that it remains a cohesive discipline, even as it advances. To this day, mathematics has not fragmented into isolated, disconnected fields; it remains unified. While the boundaries of modern mathematics continue to expand and some ideas may seem less central, eventually fading from the forefront, the fundamental principles endure.

With that, I will bring this lecture to a close. Thank you all.

Mathematics: A "Living" Discipline*

Throughout the history of human thought, mathematics has held a fundamental and unparalleled position. Spanning thousands of years—from Babylonian algebra and Greek geometry to the mathematical traditions of China, India, and the Arab world, and culminating in the remarkable advances of modern mathematics—its trajectory has occasionally been interrupted, but its objects and methods have remained consistent. The essence of mathematics lies in "numbers" and "shapes," although modern concepts have evolved far beyond their primitive meanings. Its principal tool is logical reasoning, which has constructed a robust intellectual framework. The utility of these results in other disciplines is not only predictable but has often surpassed all expectations. Mathematics has become an indispensable component of education and a cornerstone for other sciences, providing ideal models that help uncover fundamental laws.

In such an era of vibrant development, mathematics faces enormous challenges: It must simultaneously strengthen its existing foundations and adapt to external impacts. The need for a comprehensive mathematical encyclopedia is evident. Yet, the sheer magnitude of this endeavor has deterred even the most capable mathematicians.

We are fortunate to have the Soviet *Encyclopedia of Mathematics*. Its contributions to the field are immeasurable. It is crucial to recognize that mathematics is a "living" discipline—its content continually evolves and advances. Much of the mathematics studied in today's universities and research institutes did not exist

---

* The *Encyclopedia of Mathematics*, edited by the esteemed Soviet Academician I. M. Vinogradov, was first published in 1986. Shortly thereafter, a supplementary English edition was organized in the Netherlands, with contributions from 180 prominent Western mathematicians. The Chinese translation, published by the Science Press, spans five volumes, with the first volume released in 1994. The inscription was by Su Buqing, and the preface was written by Chern, which is this article.

fifty years ago, while other areas preserve the distilled brilliance of past intellectual giants, whose ideas remain as fresh and relevant as ever. I propose that the *Encyclopedia of Mathematics* include a semiannual appendix to update its content with new developments and to revise older topics. Exceptional contributions should not be constrained by editorial conventions, and the book itself should undergo periodic comprehensive revisions.

Faced with this monumental intellectual edifice, one may feel a sense of awe. The vast scope of the *Encyclopedia of Mathematics* transcends the grasp of any single individual. The role of mathematicians is to delve into selected topics, expanding and enriching them. This work will facilitate the expansion of their thinking horizons.

I firmly believe that mathematics is on the cusp of a golden age, one in which Chinese mathematicians will play a significant role. I hope this encyclopedia will serve as a meaningful role in shaping that future.

Do Good Mathematics*: A Speech at the Shanghai Mathematical Society

It is a pleasure to meet with you all, and discuss our shared goal of advancing mathematics in China. In my view, the cornerstone of this effort is cultivating China's own high-level mathematical talent.

Given the current state of affairs, training young mathematicians in China is entirely achievable. We already have a strong foundation of excellent mathematicians, and many PhD graduates from Chinese universities demonstrate academic abilities comparable to their international peers. For instance, at the NKIM, where I work, we have a PhD graduate from Jilin University whose talent is truly remarkable. I recommended him to Johannes Gutenberg University of Mainz in Germany to conduct research under Professor Kreck, an eminent topologist. After a year of outstanding work, he returned to China during this year's Spring Festival. His impressive performance has earned him two invitations for international collaborative research. We encourage him to visit more institutions.

Over the years, we have sent several young mathematicians abroad, and they are gradually returning. Although their numbers are still limited, they are beginning to make an impact. Meanwhile, economic challenges in the United States have made it difficult to secure positions in mathematics, a trend that will likely persist for some time. For those going abroad, pursuing postdoctoral research rather than another PhD is preferable. Ultimately, retaining more mathematicians in China is essential for raising the level of mathematics at our universities and research institutions.

---

* This article is a speech delivered by Professor Shiing-Shen Chern at the Shanghai Science Hall on January 6, 1994, at the invitation of the Shanghai Mathematical Society, addressing young mathematicians. The title was added by the editor of this book. It was originally published in *Mathematics Teaching*, no. 1 (1995), ed. Mathematics Department of East China Normal University.

The primary challenge facing mathematics in China today is insufficient funding. Although the government has established the Tianyuan Fund to support mathematical research, showing a commendable level of attention, the total funding remains modest. By the time resources are allocated, the amounts available locally are quite limited. I am particularly concerned about the stipends for graduate students. Could we offer special scholarships to exceptionally talented graduate students? For example, Shanghai could fund 20 outstanding students each year, providing them with a monthly stipend of 500 yuan. The total annual cost would be a mere 100,000 yuan. Many generous entrepreneurs could easily sponsor such an amount to establish dedicated scholarships. The issue lies in our outreach—potential benefactors are simply unaware of the need.

Currently, fewer students are pursuing mathematics, as many are drawn to business ventures. This trend is not unique to China; it is a global phenomenon, including in the United States. However, this is not necessarily a cause for concern. Why should those without a genuine interest in mathematics study it? If students lack true passion for the subject, it's better they choose other paths. A smaller number of highly committed students is far more effective for fostering talent.

Our efforts should focus on supporting those who have a deep love for and exceptional aptitude in mathematics. While it is important to improve living standards for mathematicians, we must be cautious about creating an overly comfortable environment. Staying in luxurious hotels in Shanghai with exquisite cuisine may not necessarily align with the spirit of mathematical exploration. For mathematicians, joy lies in something entirely different—gathering to discuss ideas, sharing insights, and experiencing the unparalleled exhilaration of a breakthrough, whether it be a simple idea or a profound discovery. Such moments are the pinnacle of fulfillment.

As a profession, mathematics offers a modest lifestyle—it is not a path to wealth—but it does have one distinct advantage: job security. Becoming a mathematician requires at least ten years of dedicated effort, and not just anyone can replace you. This makes it a relatively "safe" career. Personally, I believe there is no need for too many people to study mathematics. In a sense, the fact that "fewer people want to study mathematics" might even be an opportunity to support those who are truly dedicated to the field.

The declining number of mathematics students abroad could be China's chance to establish itself as a "mathematically advanced country." Of course, we also hope that the compensation and treatment of Chinese mathematicians will gradually align with international standards.

Next, I would like to discuss how to approach the study of mathematics. Chinese mathematicians should focus on developing their own mathematics, rather than merely following in the footsteps of others. In the early years of China's opening up, it was essential to invite some of the best international mathematicians to understand their work and appreciate their achievements. However, Chinese mathematics must have its own distinctive problems—problems that are proposed and solved by Chinese mathematicians within China. Unlike other fields, where the world often focuses on one or two major issues at the same time, mathematics offers a vast array of possibilities. We have the freedom to choose our own research topics, and they need not always be the most popular ones.

For example, when I worked on differential geometry in the past, it was considered a relatively obscure field. Some topics may seem outdated, but this is not necessarily the case. Sophus Lie's introduction of continuous group theory and his three-volume work *Theorie der Transformationsgruppen* still contains many ideas that can be further explored and developed, with significant relevance to today's research.

So, how should we choose our research topics? What makes a "good" problem, and what makes one less promising? There is no fixed method for choosing problems, and each researcher has their own criteria. Some people tend to describe their own work as the best, but this is often not the case. When I was in Hong Kong, we had the opportunity to discuss this matter. I suggested that we look at the problems posed and researched by widely recognized great mathematicians to gain some perspective.

At the turn of the 20th century, in 1900, Hilbert presented 23 mathematical problems at the International Congress of Mathematicians in Paris, which had a profound and lasting influence on the development of mathematics throughout the century. These problems touched nearly every area of 20th-century mathematics. Hilbert proposed two criteria for what makes a "good" problem. The first is clarity. He quoted Lagrange, the greatest living mathematician of the time, saying, "A mathematical theory should be able to be explained to the first person you meet on the street." Clear and understandable problems attract interest, while overly complex ones often discourage us. The second criterion is that the problem should be difficult but not unsolvable. It should not waste people's efforts.

Among the two problems Hilbert discussed, one was Fermat's Last Theorem, which states $x^n + y^n = z^n$ ($n \geq 3$) has no positive integer solutions. This problem, which had remained unsolved for centuries, was believed to have been solved recently, though it turned out that there was still a gap in the solution. Recently, it seems that this obstacle may soon be overcome. Fermat's Last Theorem sparked the development of algebraic number theory. Gauss, in his youth, wrote *Disquisitiones Arithmeticae*, which is a fundamental work in the field. His contributions were pioneering, and differential geometry also owes its foundations to him. Later, Hilbert made significant contributions to number theory, and his early work, *Zahlbericht*, remains influential even today. The second problem Hilbert discussed was the famous "three-body problem" in celestial mechanics, which involves many branches of analysis, geometry, and topology, and has important applications. Poincaré wrote two major books discussing this problem, and more recently, Professor Xiang Wuyi has offered new insights into this area. Overall, Hilbert's lecture is one that deserves close attention. In 1976, the American Mathematical Society held a special conference to discuss the progress of his 23 problems.

Speaking of good mathematical problems, I am reminded of the International Mathematical Olympiad. It is truly an honor for Chinese students to win medals in international competitions, and I have repeatedly expressed my admiration and encouragement. However, I do not consider the problems posed in these competitions to be particularly profound mathematical problems. If a problem can be solved by a student in just a few hours, it is likely lacking in deeper significance. Some problems certainly require skill to solve, but this kind of skill is not necessarily the hallmark of a strong research topic.

John Nash, the Nobel laureate in Economics this year, and a mathematician and personal friend of mine, would propose all kinds of eccentric problems. For instance, he once discovered that four cities in Europe could form a perfect square on a map. While this was an interesting discovery, it didn't carry much deeper meaning.

While I cannot precisely define what constitutes a "good" mathematical problem, I believe that one should always select topics that have greater significance and pursue them with personal vision. If we focus on important problems and introduce new concepts, I am confident we will succeed.

Finally, I want to emphasize that mathematical research should not be concentrated in one or two areas; it should be cultivated across the entire country. Often, great mathematicians emerge from smaller regions. However, these mathematicians must receive rigorous training. Many individuals who read about mathematical problems in the news and try to solve them without understanding what constitutes a proper "mathematical proof" are unlikely to make it. Serious mathematical research still requires a solid foundation.

On the Absence of a Nobel Prize in Mathematics*

After reading Mr. Huang Wenzhang's insightful article in *Communications*, vol. 21, no. 4, I found myself deeply moved and inspired. I offer a brief reflection here, perhaps for a light-hearted laugh.

▸ Why Is There No Nobel Prize for Mathematics?

The answer is simple: The Nobel Prize honors contributions that directly impact human happiness. That's why it includes categories like Peace, Medicine, and Literature. The founders were farsighted enough to foresee the massive developments in fields like Physics and Chemistry—this was truly remarkable foresight. The first prize was awarded in 1901, and the first Physics prize went to Wilhelm Roentgen for his discovery of X-rays.

Mathematics, however, serves an indirect role. Without complex numbers, there would be no electromagnetism; without Riemannian geometry, no general relativity; without the geometry of fiber bundles, no gauge theory … The profound study of material phenomena is deeply intertwined with advanced mathematics, which remains a mysterious phenomenon in scholarship.

Science requires experimentation, but experiments can never be absolutely precise. In contrast, if there is a mathematical theory, it can provide entirely correct inferences. This is why science cannot do without mathematics. Many fundamental concepts in science are expressed through mathematical frameworks. Hence, mathematicians continue to make a living, though it's only natural that they do not receive a Nobel Prize.

---

* Originally published in *Mathematical Communication* 23, no. 2 (1999).

▸ The Absence of a Nobel Prize in Mathematics: A Blessing

Mathematics is a magnificent discipline, a miracle of human thought that develops in close connection with other sciences.

One of the distinctive features of mathematics is the abundance of simple yet profoundly challenging problems. These problems can captivate a person's mind, keeping them restless for days or even years. But when the breakthrough finally comes, the joy is indescribable. Let me illustrate this with three examples:

Fermat's Last Theorem. Consider the equation, $x^n + y^n = z^n$, where $n > 0$ and x, y, z are integers. It has long been known that when $n = 2$, this equation has infinitely many solutions. However, Fermat famously claimed that for $n \geq 3$, there are no solutions unless $xyz = 0$. Despite its apparent simplicity, this conjecture defied proof for centuries. It was not until recently that Professor Andrew Wiles of Princeton University provided rigorous proof, marking one of the most celebrated achievements in the history of mathematics.

Sphere Packing Problem (Kepler's Conjecture). In n-dimensional space, how can spheres of equal diameter be packed most densely? A related, simpler question is that for a given sphere, how many others can simultaneously touch it? In $n = 2$, where spheres are circles, it is evident that 6 circles can touch a central circle—readers can verify this by drawing. In three-dimensional space, it can be proven that 12 spheres can touch a central sphere, leaving some gaps, but a 13th sphere cannot fit. This problem famously sparked a debate between Newton and Gregory. A recent elegant proof of this result can be found in the work of mathematician Xiang Wuyi.

Fundamental Theorem of Algebra. The equation, $x^n + a_1x^{n-1} + \ldots + a_n = 0$, where a_1 is a real or complex coefficient, always has at least one complex root.

These are all significant problems.

Mathematics is filled with such simple yet perplexing problems, making it a domain of endless fascination and joy. It is a serene field—a realm without grand prizes, where equality prevails. All things considered, the absence of a Nobel Prize in mathematics might actually be a blessing for the discipline.

Infusing Humanity into Mathematics for Greater Health and Joy*

Welcome, everyone, to this conference. We are gathered in an ancient land, vastly different from Western Europe, the birthplace of modern mathematics, in many respects. The year 2000 was designated as China's Mathematics Year, with the goal of inspiring more people to connect with mathematics. Today, we find ourselves with expansive fields of study and a growing community of dedicated mathematical experts. Once, mathematics was largely a solitary pursuit, but now we have built a thriving mathematical community. In this new era, one of our main tasks is to ensure that the broader public is aware of the remarkable progress we've made. Clearly, there is still much to be done in terms of outreach and popularization. I believe the challenge before us may be to craft research papers that not only delve into the complexities of the subject but also offer a historical and accessible introduction to it.

The phenomenon of the internet is nothing short of global, transcending borders. Recent research shows that different fields are no longer isolated but are increasingly interconnected, even blending together. In fact, we might soon witness the unification of pure and applied mathematics, and who knows? Perhaps a new genius, on the level of Gauss, will emerge to shape the future of the field.

China still has a long way to go in the domain of modern mathematics. However, in recent years, our performance in the International Mathematical Olympiad has been exceptional. China has laid a strong foundation and possesses an advantage in

---

* This article is the speech given by Mr. Chern at the opening ceremony of the 2002 International Congress of Mathematicians in Beijing. It was originally published in the *Chinese Mathematical Society Newsletter*, no. 3–4 (2002): p. 8.

"quantity" (the number of people). This conference, I believe, holds the potential to be a turning point, a milestone in the history of Chinese mathematics.

Confucius' teachings have shaped Chinese thought for over two millennia. His central concept, "*ren*" (仁, benevolence), which literally translates to "two people," emphasizes the importance of human relationships. Modern science, as we know, is a highly competitive field, but I believe that by infusing more human elements into mathematics, we will make our discipline not only healthier but also far more fun. Let us hope that this conference will mark the beginning of a new era in the development of mathematics.

Mathematics: Brimming with Vitality*

From the 19th to the 20th century, mathematics experienced remarkable growth. It has evolved with a sense of continuity, often centered around a unifying theme that radiates outward into diverse areas. This expansion has not only clarified foundational concepts but also led to various applications. Recently, the intertwining of mathematics with theoretical physics, the resurgence of developments in number theory, and the advent of computers have all sparked new mathematical questions, providing valuable perspectives on long-standing problems. These signs point to the undeniable vitality of mathematics, and there is much to look forward to in the 21st century.

In recent years, mathematics in China has made tremendous strides. The question now is how can we take this progress and propel it even further? Back in the 1920s, France was home to some of the most brilliant mathematical minds—such as Picard, Hadamard, and Montel. At that time, these seasoned mathematicians were focusing largely on complex analysis, which had become somewhat disconnected from modern mathematics like abstract algebra and topology. Some young French mathematicians, recognizing that they did not necessarily need to follow in the footsteps of their predecessors, decided to pursue their own studies and pave their own path. This led to the birth of the influential Bourbaki school, which played a pivotal role in shaping the course of mathematical development.

Here, I'd like to share a story: Some might wonder, "What do mathematicians spend their time on? What's the point of studying these polyhedra?" I believe, titanium compounds in chemistry are related to the concept of regular polyhedra. This is a fascinating fact—after two thousand years, regular polyhedra have found

---

* This article is an interview with Mr. Chern, conducted by Wang Jing, a reporter from *Science Times*, during the 2002 International Congress of Mathematicians in Beijing. It was first published in *Science Times* (August 20, 2002).

applications in chemistry. In fact, some mathematicians are now studying the connection between polyhedra and molecular structures. It is well-known that viruses in biology often take on polyhedral shapes. This demonstrates how what was once considered a "fantasy" by mathematicians has, over time, become highly practical.

Now, let's talk about mainstream and non-mainstream mathematics. As we know, mathematics is a highly versatile discipline. For example, it doesn't require much in the way of equipment. Thanks to email and the ease of accessing information, mathematics remains a highly personal endeavor, unlike other disciplines that heavily rely on equipment. In other fields, people compete to work in the most important directions, making their contributions in mainstream areas. Mathematics, however, is different. Since there are many directions in mathematics, and it is a personal endeavor, not everyone needs to focus on mainstream mathematics.

In 1943, I was teaching at the National Southwest Associated University. That year, I was invited to the Institute for Advanced Study of Princeton University, where Yang later became a professor. Near Princeton is a small town called New Brunswick, which houses Rutgers University. In August, I traveled to Princeton, and shortly after, I attended the American Mathematical Society's Summer Meeting in New Brunswick. Since it was nearby, I attended the talks and caught up with friends. On one occasion, I was chatting with a highly respected American mathematician who asked what I was working on. I replied, differential geometry. He immediately said, "It is dead." This was in 1943, but by the end of the war, differential geometry had become one of the mainstream areas of mathematics.

Therefore, I believe that mathematicians have the potential to discover directions that, while not mainstream today, are highly meaningful and have great promise for the future. Mainstream fields are already dominated by many of the world's leading experts, and large amounts of funding are poured into them. Competing directly with such forces may not be realistic. Instead, it may be more productive to focus on other areas that are equally significant. My hope is that Chinese mathematics can establish itself in certain fields, excel, and develop its own distinctive characteristics. This approach is not without precedent in history. For instance, before World War II, Poland focused on logic and point-set topology. Based on simple axioms, they derived numerous significant results. Finland, too, achieved notable success in complex analysis, and continues to lead the world in the extension of quasi-conformal mappings. Their success stemmed from choosing to work in areas where other countries were not involved, ultimately positioning themselves at the forefront of those fields. There are many such examples.

What is the most exciting developments in mathematics today—the current mainstream mathematics? As I mentioned earlier, I am not an advocate for everyone rushing to focus solely on mainstream mathematics, but it certainly holds great importance. Mainstream mathematics is characterized by monumental contributions—deep theorems with broad and far-reaching implications, whose proofs require extraordinary effort. If I had to select a contemporary example of such a contribution, it would undoubtedly be the Atiyah-Singer Index Theorem. Atiyah, the president of the Royal Society, visited Beijing and delivered a lecture on this subject. The index theorem can be seen as the modern extension of many of the ideas I have discussed, taking algebraic equations, Riemann surfaces, and genus theory, and expanding them from lower dimensions to higher and even infinite dimensions.

Thus, I believe that mathematical research is not just profound, difficult, and powerful, but that it remains a cohesive discipline, even as it advances. To this day, mathematics has not fragmented into isolated, disconnected fields; it remains unified. While the boundaries of modern mathematics continue to expand and some ideas may seem less central, eventually fading from the forefront, the fundamental principles endure.

The foundation of Chinese mathematics must be deeply rooted within China itself. Now, I speak on the topic of mathematics in the 21st century—specifically, how Chinese mathematics can develop and how China can position itself to have significant advantages in certain areas of mathematics in the 21st century. The solution to this challenge is simple: We must cultivate talent, identify capable individuals, and provide opportunities for young people to develop in mathematics. To elaborate further, the goal is to establish ten world-class mathematical research institutes across China. Given the vastness of China, these institutes should not be limited to Beijing but should also be spread throughout other regions.

The foundation of Chinese science must be deeply rooted within China itself as well. Only after establishing strong roots here can science and technology begin to flourish. However, attracting talented individuals to work in mathematics is no simple task. In 1984, I began the process of establishing the NKIM. Initially, I thought the solution would be straightforward—simply invite talented people to work at the institute. But due to various challenges, this proved to be more difficult than expected. To create a top-tier research institute, we need top-tier mathematicians. With top-tier mathematicians, even if the building is modest, the equipment is basic, and the resources are scarce, the institute can still be world-class. However, if the building is grand, the library is stocked with books, and expensive equipment is available, but there are no mathematicians to conduct high-level research, then none of it matters.

Recognizing this, I shifted my approach. I focused on training our own young talent, nurturing homegrown mathematicians, and sending them abroad for study. We also invited the best mathematicians from around the world to provide guidance and mentorship.

I am pleased to report that these efforts are beginning to bear fruit. For example, He Zhengxu went to the University of California, San Diego, to study under Friedman, a Fields Medalist and a prominent young leader in the field. Friedman himself has told me that He Zhengxu is his best student. I could mention many more, but time is limited.

To develop mathematics, it is essential to establish research institutes of a certain level. But what exactly defines a research institute of sufficient level?

First, we must offer advanced foundational courses. When students arrive, they need foundational training, which necessitates offering courses of the highest caliber. This "foundational training" comprises two components. The first is the development of logical reasoning skills. A student must learn to distinguish between correct and incorrect reasoning, ensuring that every step in a proof is sound. Rushing to conclusions with sloppy work leads only to errors. The second component involves a broad understanding of mathematics. A student should be able to form a comprehensive judgment of the discipline as a whole. In the past, discipline related to analysis was more important. Since the 20th century, algebra, became more prominent, followed by topology and other areas. A strong research center should be able to offer these basic courses.

Second, we must recruit exceptional students. Every year, we send top students to participate in the International Mathematical Olympiad, and many of them excel. While excelling in high school mathematics does not necessarily translate into a future in mathematical research, I hope that some of these students will pursue mathematics and achieve great things. I met with some of the award-winning students from Beijing's math competitions. I told them that pursuing mathematics would offer them a promising future. In ten or twenty years, the world will certainly face a shortage of mathematical talent. Today's young generation is less inclined to study mathematics, which will inevitably result in a talent shortage. I can understand why students may shy away from mathematics—it is difficult, and there is no guarantee of success. After years of hard study, many students still find themselves far from becoming mathematicians. Meanwhile, other fields like computer science are competing for talent. A successful career in software engineering demands great skill, but the preparatory knowledge required is much less than that needed for mathematics. Mathematicians often feel like they are constantly learning, with an endless stream of books to read. As a result, many gifted individuals are drawn to the tech industry. Some math PhDs, after graduating, choose to work in stock market, writing algorithms to predict stock market movements so that wise investment decisions are made. While they may still be employees, their salaries often far exceed those of university professors. Thus, the loss of mathematical talent is a global issue.

In contrast, China's situation is more optimistic because, even if some talent is lost, we can still cultivate more. Those who are lost and achieve financial success often return to help build the infrastructure for mathematics. Ultimately, we must adopt a particular mindset: China should become a factory for producing mathematicians. Those who leave should be encouraged to return, and if they don't, we should continue to send them abroad. China has an abundance of talent, and sending some abroad to make a global impact is worthwhile.

What we must focus on is investing modestly to build a strong foundation and offer quality courses. Once the foundation is solid, whether those who leave return will become less important. This is my preliminary vision for the future.

For example, participants in the International Mathematical Olympiad are typically outstanding in mathematics. If they enter a university mathematics department, I recommend offering them scholarships right away. Though the financial reward may be modest, the impact can be significant, and it serves as an encouragement for others. Chinese students tend to listen to their parents and teachers, and if a child shows talent in mathematics, a little persuasion from their family and mentors can often set them on the path to a mathematics career.

For gifted students in mathematics, studying abroad is just a matter of time. If they excel in mathematics domestically, going abroad becomes relatively easy. If they perform well at home, they can skip graduate school and even become professors abroad right away. China already has the conditions to produce world-class mathematicians, and we must have confidence in this.

In terms of student development, I advocate for mobility. In the 19th century, Germany was, undoubtedly, the world leader in mathematics. German university students had the freedom to register at any university. One semester, they might attend lectures by Weierstrass in Berlin, and the next, they could attend Schwarz's class in Göttingen, with complete academic freedom. Professors also enjoyed this mobility. For example, Berlin University had Planck and Einstein, and if a theoretical physicist found no prospects at Berlin, they could easily move to another institution to start fresh.

I hope that Chinese students and professors will have similar opportunities for mobility. Professors could teach at different universities for half of the year, and faculty at various mathematics institutes could exchange positions as well.

Dialogues on Mathematics

Mathematicians Must Pose Their Own Questions and Pursue Work with Developmental Potential*: A 1991 Dialogue with Zhang Dianzou

Göttingen, Hamburg, and Paris

Zhang: During the decades following World War II, differential geometry emerged as a mainstream theme in the development of mathematics, often regarded as a "hot topic." What inspired you to choose geometry as your field of study?

Chern: When it comes to "hot" topics, I've never been one to chase trends. My decision to delve into geometry was entirely shaped by my circumstances. At Nankai University, I encountered Professor Jiang Lifu, who specialized in geometry. After graduating, I met Professor Sun Guangyuan, who was also working in the same field. These experiences determined my

---

* On October 28, 1991, Zhang Dianzou visited the office of Mr. Chern, then Director of the Mathematical Sciences Research Institute. Nestled on the third floor, the office offered a stunning view of the Golden Gate Bridge and San Francisco Bay. Against this picturesque backdrop, Zhang began an enlightening conversation with the mathematical master.

trajectory in differential geometry. However, if we're talking about personal inclinations, I might have been more drawn to algebra.

Zhang: Was differential geometry considered "hot" during the 1930s?

Chern: Not particularly. Analysis has always been the mainstream of mathematics. At that time, the Göttingen school in Germany, led by Courant in analysis and Emmy Noether in algebra, was highly influential. In Britain, Hardy and Littlewood dominated the landscape of function theory—analytic number theory. In France, function theory was flourishing under the leadership of luminaries like Picard, Lebesgue, and Montel. Meanwhile, functional analysis, championed by Hilbert and Banach, was experiencing great popularity. Although Riemannian geometry gained some momentum with the rise of general relativity, it remained a niche area—a realm for the few and far from being a "hot" topic.

Zhang: Before World War II, Göttingen was considered the global center for teaching mathematics. Why didn't you go to Göttingen to "pay homage" and instead chose Hamburg?

Chern: I believe choosing where to work should align with one's personal goals. While meeting prominent figures can give you stories to tell, it is not essential for academic growth. That said, mathematical hubs with renowned figures naturally attract talented individuals, fostering a unique and stimulating environment.

I chose Hamburg primarily because Blaschke had visited China to lecture. I was able to understand his material (around the same time, G. D. Birkhoff from the US also gave lectures, but I couldn't follow his talks), which allowed for in-depth discussions. Moreover, I had read papers from Hamburg University's journal that captivated me. These factors ultimately led me to Hamburg.

Zhang: Were there prominent geometers in Göttingen at that time?

Chern: Yes, there were. For example, S. Cohn-Vossen, as well as Herglotz, who was a remarkable mathematician with a broad range of work. Herglotz also worked on geometry, and rigid body geometry includes the celebrated Herglotz Theorem. Still, I felt Hamburg was a better fit for me.

Zhang: Did you gain much during your time in Hamburg?

Chern: Certainly. Apart from Wilhelm Blaschke, Emil Artin and Erich Hecke were exceptionally strong mathematicians. Among the younger mathematicians, there were Erich Kähler, Helmut Petersen, and Hans Zassenhaus. Kähler, in particular, was of great help to me.

I remember that during 1934–1935, much of my time and energy went into Kähler's seminar. The discussions focused on a famous monograph, Introduction to *Systems of Partial Differential Equations*, which later became known as the Cartan-Kähler Theorem. The seminar started with a packed room, including luminaries like Blaschke, Artin, and Hecke, but attendance dwindled over time. I was one of the few who stayed until the very end. My doctoral thesis combined Kähler's theories on network geometry with some earlier results. I've always approached mathematics with my own ideas, avoiding trends and focusing on what suited me best.

Zhang: In the 1930s, many students studying abroad stopped research after earning their doctorate. Yet you pursued postdoctoral work under Cartan, a leading mathematician. This shows your ambition in mathematics.

Chern: My motivation for studying mathematics wasn't ambition but rather a desire to deeply understand the subject. If one's goal is fame or wealth, mathematics is certainly not a shortcut. At that time, Cartan was the foremost geometer, though his work in the 1930s was considered "ahead of its time" and not widely understood. I was deeply impressed by Cartan's profound knowledge and went to Paris in October 1936, staying for ten months.

That said, research is a demanding endeavor that does not guarantee recognition, so it is natural for many to stop once they finish their studies. This is true worldwide. At the Poincaré Institute in Paris, there were shelves of meticulously bound doctoral theses, yet most authors had long disappeared from the academic scene. Delving into mathematics is a demanding pursuit. Why do some persist? The reasons vary. For me, it was the sense of fulfillment it provided. A friend of mine, Yang Wuzhi, once gifted me a poem describing this feeling: "Solitary steps ascend a tower a hundred fathoms high." This aptly captures the mindset. Over the years, I've written numerous papers, with all the accompanying joys and struggles—it's not something easily summed up in a few words.

Zhang: Cartan's work is said to be challenging to grasp, yet you thoroughly mastered his methods. This must have been a critical step in your success.

Chern: Indeed. Cartan's contributions were primarily in two areas: Lie groups and exterior differential systems. These provided powerful new tools for differential geometry. Mastery of such techniques is essential for doing the "best" mathematics. Cartan would often assign me what he considered "small" problems. I'd return after investigating them and turn them into published papers. Perhaps because I consistently responded to his guidance, he allowed me the rare privilege of visiting his home every fortnight or so. After our discussions, I would often receive follow-up letters the next day, elaborating on the previous conversation. These interactions gave me a profound understanding of the "best" geometry of the time.

Zhang: Has Cartan's work become widely understood today?

Chern: Not entirely. Even now, many of his ideas have not been widely accepted. I've said that many modern books on differential geometry are poorly written. They often begin with the formula

$$\nabla_Z \nabla_Y X - \nabla_Y \nabla_Z X - \nabla_{[Y,Z]} X = R(Y,Z)X$$

and attempt to derive all of differential geometry from it. However, Cartan's methods incorporate two additional layers: the relationship between the connection and ω, and between the curvature and Ω. Starting solely from the above equation fails to yield significant results. Using my understanding of Cartan's approach, on the other hand, leads to many insights with ease. I have pointed this out before, but many persist with their ways. This isn't entirely bad—it allows me to keep a "secret weapon." When others fail to achieve results, I can often succeed, giving me a slight edge. Such discrepancies arise because geometry is governed by algebraic principles, and different algebraic methods produce different geometric outcomes.

Princeton and Global Differential Geometry

Zhang: Would you say that "global differential geometry" represents the "best" of the mathematics you wanted to pursue?

Chern: The evolution of differential geometry toward a global perspective is a natural progression. Once local properties are understood, it's only logical to explore their global implications. What's unexpected is the discovery of geometric phenomena with profound global significance. During my time at Princeton from 1943 to 1945, my understanding of mathematics deepened significantly. To study global geometry effectively, one requires a solid foundation in classical geometry, mastery of the then-nascent field of algebraic topology, and the ability to adapt and refine Cartan's geometric methods. This is the way to innovate and distinguish myself. Achieving this requires immense effort, and few ventured into this field. Yet, this challenge was precisely what I sought.

Zhang: You were invited to Princeton by the American mathematicians Veblen and the renowned Weyl. Did they want you to work on global differential geometry?

Chern: Not at all. My research was entirely self-directed. My trip to Princeton stemmed primarily from my correspondence with Veblen. In 1936, while I was in Paris, Veblen wrote to Cartan inquiring about projective normal coordinates, which were central to the "geometry paths" of the Veblen school. Their goal was to develop a more advanced theory of differential geometry that extended beyond the limitations of Einstein's Lorentzian space to make a unified field theory. I know that while projective normal coordinates could be defined in various ways, each method had limitations. I proposed a definition rooted in Cartan's geometric methods and sent it to Veblen. He was deeply impressed. Later, during my time at National Southwest Associated University, I sent some of my and Wang Xianzhong's papers through Veblen. This strengthened our connection, even though we had never met. Veblen was unaware of my interest in global differential geometry.

Zhang: Being invited to Princeton during the war years must have been quite rare.

Chern: Indeed. At the time, Princeton faced funding constraints, and the ongoing war made it challenging to invite researchers. They were interested in my work, but being a Chinese scholar may have also played a role—few Chinese were engaged in scientific research, and I posed no threat to American job opportunities, which might have influenced their decision on inviting Chinese first for a visit. Looking back, I was fortunate. My "best" work was completed during my time at Princeton.

Zhang: If you were fortunate to be invited to Princeton, would you say that good fortune plays a role in academic pursuits? For instance, choosing a research direction might sometimes lead to breakthroughs and other times to a dead end.

Chern: While luck has its role, I believe success primarily depends on ability. Finding oil in a vast desert isn't purely a matter of chance—it relies on geologists' accumulated knowledge and scientific judgment. Similarly, while mathematical problems exist, not everyone can solve them. Exceptional mathematicians can tackle problems that others cannot.

Zhang: Could you provide an example from global differential geometry? Why did you choose to study this field?

Chern: A mathematician must distinguish between good mathematics and bad mathematics, or less meaningful mathematics. It's like appreciating poetry or art: certain masterpieces inspire endless admiration and evoke profound respect. It is the same for mathematics. The transition from local differential geometry to a global perspective is a natural step.

But it requires thorough preparation. I recognized the importance of algebraic topology early on. In 1932, Blaschke delivered a lecture in Beijing titled "Topological Problems in Differential Geometry," though it actually addressed local differential geometry. In 1933–1934, Emanuel Sperner visited China and rigorously proved the topological Jordan Curve Theorem, which states that a closed curve divides the plane into two regions. I also attended a topology course by Jiang Zehan, though I felt I hadn't fully entered the field at that time. The situation changed with the publication of *Topology* by Alexandrov and Hopf. Algebraic topology became an indispensable branch of mathematics, and I found it deeply fascinating.

Zhang: Why, then, did you choose the Gauss-Bonnet Theorem as your starting point for research?

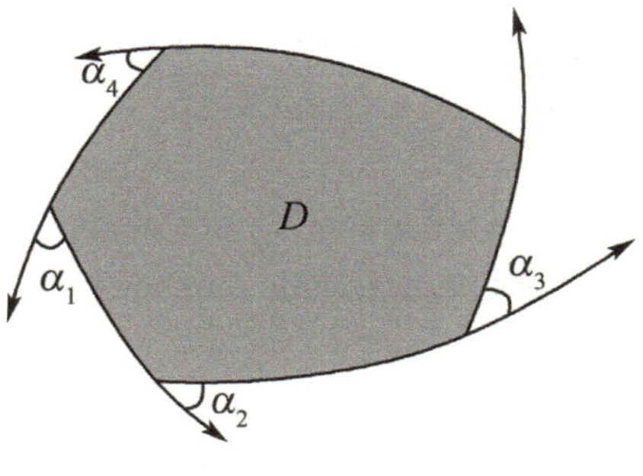

The two-dimensional Gauss-Bonnet Theorem

For a simple closed curve C on a curved surface, composed of finitely many connected curve segments, which enclose a region D, and with external angles α_i (i = 1, 2, …, m), the following holds

$$\int_C \rho \mathrm{d}s + \sum_{i=1}^{m} \alpha_i + \iint_D K \mathrm{d}\sigma = 2\pi$$

Here, ρ represents the curvature of C, and K is the Gaussian curvature. A special case of this is that the sum of the three external angles of a triangle is 2π.

Chern: When I was teaching at National Southwest Associated University, I developed a unique and unconventional understanding on this topic. The renowned topologist Hopf, in his 1925 doctoral dissertation, studied higher-dimensional extensions of the Gauss-Bonnet Theorem. He once predicted, "Extending the Gauss-Bonnet Theorem to higher dimensions is both the most important and the most challenging problem." Hopf extended the formula to the case of hypersurfaces, and Weyl also made significant contributions. Later, C. Allendoerfer and A. Weil (1906–1998) proved the general Gauss-Bonnet Theorem for higher-dimensional Riemannian manifolds. However, their approach relied on manifolds embedded in Euclidean space.

When I arrived at Princeton, I provided a completely "intrinsic" proof of the formula. My approach utilized unit tangent vector bundles, in contrast to Weyl, Allendoerfer, and Weil, who worked with a "non-intrinsic" sphere bundle. This fundamental difference led to the full resolution of the Gauss-Bonnet formula. Taking such a different and innovative path requires not only effort but also ability.

Zhang: Your most famous work is the "Chern class." Why do you think it is more significant than other characteristic classes?

Chern: To understand this, one needs a certain vision for analysis. The main types of characteristic classes are Whitney class—general topological invariants, Pontryagin class—topological invariants on real manifolds, and Chern class—topological invariants on complex vector bundles.

The key lies in the fact that I worked on complex vector bundles. Complex numbers belong to a magical realm. For instance, within complex numbers, every algebraic equation has a solution, which is not always the case with real numbers. Moreover, my research emphasized vector bundles, which describe not only the base space but also the fiber structure. This dual focus gave the "Chern class" a broader and deeper significance. These mathematical structures, which hold both local and global meaning, possess universal value and influence the entire field of mathematics. My attention was directed toward the "complex" structure. Compared to real bundles, complex bundles are simpler. Algebraically, the complex number field exhibits simpler properties, and in group theory, complex linear groups behave similarly. These factors likely explain why complex vector bundles play such an important role.

Mathematicians and Mathematical Schools

Zhang: Everyone wants to improve their abilities. How can one achieve this? Is diligence the key?

Chern: Diligence is, of course, necessary, but it cannot be measured by appearances alone. Some people spend their days in the office, tirelessly reading, calculating, and filling bags with drafts. That is one form of diligence. Others, however, might stroll around, observe, or chat with colleagues, and their seemingly casual efforts are also diligence and may lead to even greater achievements. For example, in Göttingen, van der Waerden spent his days immersed in the office, while Cohn-Vossen moved about, exploring and engaging. Both achieved remarkable results. Cohn-Vossen's work on two-dimensional manifolds, for instance, was groundbreaking. Even while wandering, he was deeply engaged in thought.

Zhang: Mathematicians often appear to be busy craftsmen or artisans, constantly calculating, solving problems, honing techniques, and proving conjectures.

Chern: Craftsmen and artisans are indispensable, and exceptional craftsmanship can be immensely valuable. However, designing the grand structure of mathematics requires mathematicians. The creation of new disciplines often hinges on innovative mathematical concepts and ideas. These cannot be achieved solely by practicing techniques in an office. To succeed, one must explore broadly, engage in discussions, synthesize knowledge, and expand their horizons.

Zhang: From our conversation, I sense that you place great importance on framing questions, anticipating the next steps, and observing future directions.

Chern: Precisely. I believe that mathematicians should aim to work on problems that will shape future developments, not just focus on immediate outcomes. Anticipating the future doesn't mean drafting detailed plans on paper; it's about contemplating the right direction. With a clear sense of direction, you can identify your own problems and develop your own ideas. Solving conjectures posed by others is commendable and significant. However, solving profound theoretical problems of your own making is even more rewarding and impactful. When I was at Princeton, I often had casual conversations with the great mathematician Weyl. He always had a forward-looking perspective. On one occasion, he said to me, "It seems algebraic geometry will experience significant growth." Time later proved his prediction correct.

Zhang: Nowadays, many believe that the "powerful American school of differential geometry" owes much to your influence. How did you achieve this?

Chern: Academic influence primarily depends on one's work, but personality also plays a role. I enjoy engaging with others. For instance, my friendship with André Weil has spanned over half a century. I've also collaborated on papers with renowned mathematicians like Raoul Bott, Louis Nirenberg, Claude Chevalley, Phillip Griffiths, Jean-Pierre Serre, and Friedrich Hirzebruch. In addition, I've mentored over 40 students who earned their PhDs under my supervision and worked with many young mathematicians, coauthoring papers. I believe I have a good eye for identifying intriguing problems worth pursuing.

Zhang: Some mathematicians, however, are less social—for example, Wei-Liang Chow.

Chern: Mr. Chow is an old friend. When he married Margot Victor, I was the only Chinese guest at their wedding. Chow worked at night and slept until two or three in the afternoon. Since German banks closed at 1 p.m., he often relied on me to withdraw money for him. Chow made significant contributions to algebraic geometry, but he is naturally reserved, preferring a quiet family life. During the selection of academicians for the Academia Sinica, many outside the field did not fully understand his work. And I had to say, "If Chow isn't chosen, we who are selected should feel ashamed." He was eventually elected but never participated in any activities.

Zhang: Do you think mathematicians should take on public roles or administrative responsibilities to expand their influence?

Chern: No, not at all. I'm not fond of administrative duties myself; I even declined the presidency of the American Mathematical Society. However, pioneering initiatives like founding these research institutes are meaningful. Here, let me share an anecdote from Chinese mathematical history. The Chinese Mathematical Society wasn't established until 1935 because northern mathematicians like Jiang Lifu and Feng Zuxun avoided administrative burdens. Eventually, Gu Cheng from the south took on this task but felt unqualified, so he invited Hu Dunfu from Jiaotong University to be the first president, and the society was finally founded in Shanghai. During the war, Gu joined the Wang Jingwei regime, and a new mathematics society was formed in the rear areas with Jiang Lifu as president. After the war, the two societies merged, with Jiang as president. Hu was pleased, and everyone worked together harmoniously.

Zhang: Can you share your exchange with the French Bourbaki school?

Chern: André Weil, the soul of the Bourbaki group, is a close friend of mine. Other members like Henri Cartan, Jean Dieudonné, and Claude Chevalley are also my friends, though I never joined their activities. From 1936 to

1937, I was a postdoctoral researcher under Elie Cartan at the University of Paris. At that time, the early Bourbaki group held semiweekly "Julia" seminars focusing on Elie Cartan's work. I interacted with them during this period.

Zhang: The Bourbaki school is best known for its *Eléments de mathématique* series. What is your opinion of it?

Chern: According to Weil, in the 1930s, they felt that many mathematicians' work lacked rigorous logical foundations. To avoid perpetuating errors, they began writing a series of books starting from basic set theory, ensuring that everything was reliable. These are foundational books, not textbooks.

Zhang: Some mathematicians criticize Bourbaki's approach for stifling creativity.

Chern: That's the reader's issue. Authors present their perspectives; you can choose to follow them or not. The responsibility doesn't lie with the authors. In fact, Weil and others produced profound and expansive mathematical work and didn't limit themselves to the style of *Eléments de mathématique*.

Zhang: What influence has Bourbaki had on geometry?

Chern: Not much. For instance, in differential geometry, consider the Stokes Theorem. What exact conditions are necessary for it to hold? Smoothness is a sufficient condition, but what degree of non-smoothness can still allow the theorem to work? That's hard to pin down. The Bourbaki group eventually realized that certain mathematical structures can't be made entirely clear-cut. As a result, their series was never completed as originally envisioned, and their idea of publishing comprehensive volumes on each topic was abandoned. Additionally, their neglect of applied mathematics was a mistake.

Zhang: Dieudonné's advocacy for "overthrowing Euclidean geometry" in high schools caused considerable backlash.

Chern: I don't agree with Dieudonné on that. Their series represents just one aspect of mathematics and shouldn't be taken as a model for everything. If there were only one framework, mathematics would lose its vitality.

Mathematics in the Soviet Union and China

Zhang: Could you share your thoughts on geometry research in the Soviet Union?

Chern: I'd like to focus only on Soviet contributions to topology. Pioneering mathematicians like Alexandrov, Kolmogorov, and Pontryagin made outstanding advancements in the field. The 1935 Moscow Topology Conference served as a significant showcase of their achievements. Later, Pontryagin shifted his focus to cybernetics. Alexandrov, with his strong preference for closed sets, seemed somewhat one-sided in his approach. If you read Whitney's papers carefully—such as those in the American Mathematical Society's centennial commemorative volume—you'll see that his work after 1935 became a cornerstone for the subsequent development of topology. Whitney was a giant in mathematics, though a quiet and dedicated scholar known mainly to mathematicians. Nevertheless, he received the Wolf Prize for his contributions.

Zhang: Do you have any thoughts on the study of the history of mathematics?

Chern: I do have some ideas. Many current works on the history of mathematics resemble "news compilations," recording events such as awards or conferences without delving into the pivotal developments in mathematics itself. Some have suggested I write a history of differential geometry, and I've considered it since I actively contributed to its progress during certain periods. However, my current research workload is substantial, and it's uncertain when I'll have the time to begin.

I also have lingering questions about Chinese mathematical history. During the Song (960–1279) and Yuan (1206–1368) dynasties, Chinese mathematics advanced rapidly. Was this due to foreign influences, such as from Arab scholars? Qin Jiushao, in the preface to his *Shushu Jiuzhang*, mentioned receiving guidance from "strangers." What did he mean by that? Did Chinese mathematicians collaborate with one another? Were there institutions dedicated to teaching mathematics at the time? These are fascinating questions worth exploring.

Zhang: This year marks your 80th birthday. Many have extended their congratulations and best wishes for your continued health.

Chern: Reaching 80 isn't particularly something to be happy about. The future belongs to the young, and our hopes rest with them. At my age, physical activities are no longer feasible, and listening to music feels like a waste of time. However, my mind hasn't slowed down, so I still manage to publish a few papers each year.

Zhang: What is your outlook on the future of mathematics in China?

Chern: I'm optimistic. Young talent is emerging rapidly, both on the mainland and in Taiwan. Taiwan now has over 200 PhD holders in mathematics, and the number on the mainland is also growing quickly. Leaders in the field will naturally arise over time.

Zhang: Chinese mathematicians have yet to deliver an hour-long presentation at the International Congress of Mathematicians, and even 45-minute talks are rare. Is this due to a lack of ability or recognition?

Chern: I believe it's primarily due to a lack of recognition. For a long time, China was not part of the International Mathematical Union, so others were unfamiliar with our work, resulting in fewer opportunities to present. However, not being invited to speak isn't particularly important—it's just news that fades quickly. What truly matters is focusing on improving the quality of our work. Many exceptional mathematicians have never presented at the ICM, yet this doesn't diminish their academic standing.

Zhang: Do you think China can achieve its aspiration of becoming a "mathematically advanced country" in the 21st century?

Chern: Being a "mathematically advanced country" doesn't mean dominating the world or outcompeting everyone. It simply means establishing a strong mathematical community within China, capable of engaging with international mathematicians on equal and independent terms. Given China's size and population, achieving this should not be an issue.

Postscript by Shiing-Shen Chern (January 1992): After reading Mr. Zhang Dianzhou's interview transcript, I feel humbled. Some of my comments may reflect personal biases, so I hope readers won't take them too seriously. However, I believe the facts I presented are accurate.

The 21st century will witness a flourishing of science that will reshape the world in ways we cannot yet predict. Mathematics will undoubtedly remain a fundamental discipline because of its logical and robust foundation. Its utility across many sciences, however, retains a touch of mystery. Personally, I believe that the universe's most remarkable phenomena are rare and tend toward unity. Yet the endless depths of knowledge—such as three-dimensional geometry—remain enigmatic.

The purpose of a mathematician is to understand mathematics. Historically, the progress of mathematics follows two main paths: deepening the understanding of existing material and expanding its scope. Recent rapid developments in mathematics are dazzling, forcing mathematicians to concentrate on specific areas. Even within a small domain, one can find infinite joy. As Tao Yuanming said, "Every time I gain insight, I forget my hunger with delight." Similarly, Du Fu remarked, "Writing is a matter of eternity, / Only the heart knows its gains and losses." These sentiments encapsulate the mathematician's ultimate pursuit.

Human energy is limited. I believe mathematicians should aim to "master one field first." If they have the capacity, they can expand from that foundation to explore broader horizons. But even mastering a single field is a significant achievement.

The 20th century laid the foundation for modern mathematics in China, yielding remarkable accomplishments. The 21st century will undoubtedly be a bright era for Chinese mathematics. I hope my peers approach the future with confidence and forge ahead with determination.

A Good Mathematician Should Be a Problem Solver*: A Conversation with Jackson in 1998

Shiing-Shen Chern is one of the greatest living geometers. He was born on October 28, 1911, in Jiaxing, China. His father had a degree in law and worked for the government. When Chern was a youngster, China was just starting to establish Western-style colleges and universities. He entered Nankai University at the age of fifteen and was drawn to physics but, finding himself clumsy with experimental work, eventually settled on mathematics. In 1930 he entered the graduate school of Tsinghua University, where there were a number of Chinese mathematicians who had obtained PhDs in the West. Among these was Sun Guangyuan Sun, who had been a student of E. P. Lane at the University of Chicago. Some twenty years later, Chern became Lane's successor at Chicago. In 1932 Wilhelm Blaschke, a mathematician from the University of Hamburg, visited Tsinghua, and his lectures had a great influence on Chern.

Jackson: After your studies in China, you decided to get a doctorate in the West.

Chern: I was given a fellowship to come to the West by Tsinghua University in 1934, after one year of assistantship and three years in the graduate

* This article was originally published in *Notices of American Mathematical Society* 45, no. 7 (1998), under the title "Interview with Shiing-Shen Chern," with the interviewer being Allyn Jackson. The Chinese translation is available in *Mathematics Translation* 18, no. 2 (1999), trans. Tian Yimei and ed. Li Yezhou.

school. I decided Europe was a better place than the United States. The normal thing to do was to come to the United States, but I was not interested in Princeton or Harvard.

Jackson: Why not?

Chern: Not so good. I wanted to be a geometer. The United States did not have the type of geometry I wanted to work on, so I went to Europe. At that time, I think I had the advantage that although I was a beginning student, I had some ideas about what I wanted, about the mathematical situation in the world, who are the good mathematicians, where are the best centers. My evaluation could have been wrong, but I had my ideas. And I decided to go to Hamburg. In fact, it was a very good choice. At the end of the nineteenth century the center of science was Germany, including mathematics. The center of mathematics in Germany was Göttingen, with Berlin and Munich not far behind. And Paris, of course, was always a center.

I graduated from Tsinghua Graduate School in 1934. In 1933 Hitler took power in Germany, and there was great movement in German universities. The Jewish professors were removed, and so on, and Göttingen collapsed. And Hamburg became a very good place. Hamburg was a new university founded after the World War I. It was not so distinguished, but the math department was excellent. So I went there at the right time.

It was in Hamburg that Chern first came into contact with the work of Elie Cartan, which had a profound influence on Chern's approach to mathematics. At that time, Erich Kähler, a privatdozent in Hamburg, was one of the main proponents of Cartan's ideas. Kähler had just written a book, the main theorem of which is now known as the Cartan-Kähler Theorem, and he organized a seminar in Hamburg. On the first day of the seminar all of the full professors—Blaschke, Emil Artin, and Erich Hecke—attended.

Chern: The seminar looked like a kind of celebration. The classroom was filled, and the book had just come out. Kähler came in with a pile of the books and gave everybody a copy. But the subject was difficult, so after a number of times, people didn't come anymore. I think I was essentially the only one who stayed till the end. I think I stayed till the end because I followed the subject. Not only that, I was writing a thesis applying the methods to another problem, so the seminar was of great importance to me. I even came to see Herr Kähler after the seminar. A lot of times we had lunch together. There was a restaurant near the institute, and we had lunch together and talked about all kinds of things. My German was very limited, and Herr Kähler did not speak English at that time. Anyway, we got along. So, as a result, I finished my thesis very quickly.

Everybody knew that Elie Cartan was the greatest differential geometer. But his writings were very difficult. One reason is that he uses the so-called exterior differential calculus. In our subject of differential geometry, where you talk about manifolds, one difficulty is that the geometry is described by coordinates, but the coordinates do not have meaning. They are allowed to undergo transformation. In order to handle this kind of situation, an important tool is the so-called tensor analysis, or Ricci calculus, which was new to mathematicians. In mathematics, you have a function; you write down the function, you calculate, or you add, or you multiply, or you can differentiate. You have something very concrete. In geometry, the geometric situation is described by numbers, but you can change your numbers arbitrarily. So to handle this, you need the Ricci calculus.

Chern had a three-year fellowship, but finished his degree after only two years. For the third year, Blaschke arranged for Chern to go to Paris to work with Cartan. Chern did not understand much French, and Cartan spoke only French. On their first meeting, Cartan gave Chern two problems to do. After some time they happened to meet on the stairs at the Institut Henri Poincaré, and Chern told Cartan he had been unable to do the problems. Cartan asked Chern to come to his office to discuss them. Chern thereafter came regularly to Cartan's office hours, which often attracted a large number of visitors who wanted to meet with the famous mathematician. After a few months, Cartan invited Chern to meet with him at his home.

Chern: Usually the day after meeting with Cartan, I would get a letter from him. He would say, "After you left, I thought more about your questions ..." He had some results, and some more questions, and so on. He knew all these papers on simple Lie groups, Lie algebras, all by heart. When you saw him on the street, when a certain issue would come up, he would pull out some old envelope and write something and give you the answer. Sometimes it took me hours or even days to get the same answer. I saw him about once every two weeks, and clearly I had to work very hard. This lasted for a year, till 1937, and then I went back to China.

When he returned to China, Chern became a professor of mathematics at Tsinghua, but the War of Resistance Against Japanese Aggression severely limited his contact with mathematicians outside China. He wrote to Cartan about his situation, and Cartan sent a box of his reprints, including some old papers. Chern spent a great deal of time reading and thinking about them. Despite his isolation Chern continued to publish, and his papers attracted international attention. In 1943 he received an invitation from Oswald Veblen to come to the Institute for Advanced Study in Princeton. Because of the war it took Chern a week to reach the United States by military aircraft. During his two years at the institute, Chern completed his proof of the generalized Gauss-Bonnet Theorem, which expresses the Euler characteristic of a closed Riemannian manifold of arbitrary dimension as a certain integral of curvature terms over the manifold. The theorem's marriage of the local geometry to global topological invariants represents a deep theme in much of Chern's work.

Jackson: What among your mathematical works do you consider the most important?

Chern: I think the differential geometry of fiber spaces. You see, mathematics goes in two different directions. One is the general theory. For instance, everybody has to study point-set topology, everybody has to study some algebra, so they get a general foundation, a general theory that covers almost all mathematics. And then there are certain topics which are special, but they play such an important role in the application of mathematics that you have to know them very well, such as the general linear group, or even the unitary group. They come out everywhere, whether you do physics or number theory. So there is a general theory, which contains certain beautiful things. And the fiber space is one of these. You have a space whose fibers are very simple, are classical spaces, but they are put together in a certain way. And that is a really fundamental concept. Now, in fiber spaces the notion of a connection becomes important, and that's where my work comes in. Usually the best mathematical work combines some theory with some very special problems. The special problems call for development in the general theory. And I used this idea to give the first proof of the Gauss-Bonnet formula. The Gauss-Bonnet formula is one of the important, fundamental formulas, not only in differential geometry, but in the whole of mathematics. Before I came to Princeton in 1943 I had thought about it, so the development in Princeton was in a sense very natural. I came to Princeton and I met André Weil. He had just published his paper with Allendoerfer.[1] Weil and I became good friends, so we naturally discussed the Gauss-Bonnet formula. And then I got my proof. I think this is one of my best works because it solved an important, fundamental, classical problem, and the ideas were very new. And to carry out the ideas you need some technical ingenuity. It's not trivial. It's not something where once you have the ideas you can carry it out. It's subtle. So I think it's a very good piece of work.

Jackson: One of your other most important works was the development of characteristic classes.

Chern: The characteristic classes—they are not that impressive. Characteristic classes are very important, because these are the fundamental invariants of fiber spaces. Fiber spaces are very important; therefore, the characteristic classes come up. But it did not take me that much thought. They come up often, even the characteristic class c_1, because in electricity and magnetism, you need the notion of complex line bundles. And

1 This paper presented a proof of Gauss-Bonnet that was less desirable than Chern's in that it relied on the fact that a Reimannian manifold can be locally isometrically embedded in a Euclidean space. Chern's proof used only intrinsic properties of the manifold.

the complex line bundles lead to c_1, which comes up in Dirac's paper on quantum electrodynamics. Of course, Dirac did not call it c_1. When c_1 is not zero, that's related to the so-called monopole. So characteristic classes are important in the sense that they come up naturally in concrete problems, fundamental problems.

Jackson: When you first developed the theory of Chern classes in the 1940s, were you aware of Pontryagin's work and the fact that the Pontryagin classes of a real bundle could be recovered from the Chern classes of its complexification?

Chern: My main idea is that you should do topology or global geometry in the complex case. The complex case has more structure and is in many ways simpler than the real case. So I introduced the complex Chern classes. I read the Pontryagin papers, but the real case is much more complicated. I didn't see his full papers, but I think he made some kind of announcement in *Doklady*[2] in English. I learned from Hirzebruch the relations between the Chern classes and the Pontryagin classes. Chern classes can be expressed in terms of the curvature, in terms of the local invariant. I was mainly interested in the relations between local properties and global properties. When you study spaces, what you can measure are the local properties. It's very remarkable that some local properties are related to global properties. The simplest case of the Gauss-Bonnet formula is that the sum of the angles of a triangle is 180 degrees. It shows up already in very simple cases.

Jackson: You are seen as one of the main exponents of global differential geometry. Like Cartan you have worked with differential forms and connections and so on. But the German school, of which Klingenberg is one of the exponents, does global geometry in a different way. They don't like to use differential forms, they argue with geodesics and comparison theorems, and so on. How do you see this difference?

---

2 A journal of the Russian Academy of Sciences.

Chern: There is no essential difference. It's a historical development. In order to do, say, geometry on manifolds, the standard technique was Ricci calculus. The fundamental problem was the form problem, which was solved by Lipschitz and Christoffel, particularly Christoffel. The Christoffel idea went to Ricci, and Ricci has his book on the Ricci calculus. So all these people, including Hermann Weyl, learned mathematics through the Ricci calculus. Tensor analysis had such an important role, so everybody learned it. That's how everybody started in differential geometry, with tensor analysis. But somehow in certain aspects, the differential forms should come in. I usually like to say that vector fields is like a man, and differential forms is like a woman. Society must have two sexes. If you only have one, it's not enough.

After spending 1943–1945 at the Institute in Princeton, Chern returned to China for two years, where he helped to build the Institute of Mathematics at the Academia Sinica. In 1949 he became a professor of mathematics at the University of Chicago and in 1960 moved to the University of California, Berkeley. After his retirement in 1979 he continued to be active, and in particular helped launch the MSRI in Berkeley, serving as its first director from 1981 to 1984. Chern has had 41 PhD students. This number does not count the many students he has had contact with on his frequent visits to China. Because of the Cultural Revolution in China, the country lost many talented mathematicians and the tradition of mathematical research almost died out. Chern did much to regenerate this tradition. In particular, he was instrumental in starting the NKIM in Tianjin, China, in 1985.

Jackson: How often do you get back to China?

Chern: In recent years I go back every year, usually staying one month or longer. I started the institute at Nankai, and the most important thing is to get some good young people who will stay in China. In this respect we have been successful. Our new faculty includes Long Yimin (dynamical systems), William Chen (discrete mathematics), Zhang Weiping (index theory), and Fang Fuquan (differential topology). There are other very good young people. I think the main obstruction to the progress of mathematics in China is the low pay. By the way, the International Mathematical Union has chosen Beijing for the next International Congress of Mathematicians.

Jackson: Do you think that will be a big boost for mathematics in China?

Chern: Oh, yes. But I think what I am worried about is that there will be too many mathematicians in China.

Jackson: It's a large country; maybe they need a lot of mathematicians.

Chern: I think they don't need too many mathematicians. China is a large country, so naturally it has a lot of talent, particularly in the smaller places. For instance, there's the International Mathematical Olympiad for high school students, and China generally does very well. In order to achieve well in competitions like this, the students need training, and as a result other topics could be ignored. Now the parents in China want their children to know more English, go into business, and make more money. And these exams don't give money. One year I think they just did less of this training, and China immediately dropped. What do you do for a country with 1.2 billion people? It means that the standard of living cannot be very high, if you have any social justice.

In 1934, when Chern chose to go to Germany for graduate study, geometry was a peripheral subject in the United States. By the time of his retirement in 1979, geometry counted as one of the most vibrant specialties on the US mathematical scene. Much of the credit for this transformation goes to Chern. Still, he is modest about his achievements.

Chern: I don't think I have big views. I only have small problems. In mathematics a lot of concepts and new ideas come in, and you just ask some questions, you try to get some simple answer, and you want to give some proof.

Jackson: That's how you get your ideas, just observing things?

Chern: Yes, in most cases you don't have an idea. And in even more cases your ideas don't work.

Jackson: You're describing yourself as a problem solver, rather than somebody who builds a theory.

Chern: I think the difference is small. Every good mathematician has to be a problem solver. If you are not a problem solver, you only have vague ideas, how can you make a good contribution? You solve some problems, you use some concepts, and the merit of mathematical contributions, you probably have to wait. You can only see it in the future. It is very difficult to evaluate a mathematician or a part of mathematics. Like the concept of differentiability. Some time ago, twenty or thirty years ago, a lot of people just didn't like differentiability. I heard a lot of people who told me personally, "I'm not interested in any mathematics with a notion of differentiability." These are the people who tried to make mathematics simple. If you exclude notions involving differentiability you could exclude a lot of mathematics. But it's not enough. Newton and Leibniz, they should play a role. But it's interesting because there are ideas in mathematics which are controversial.

Jackson: Can you give some examples of controversial ideas in mathematics?

Chern: One thing is that some of the papers nowadays are too long. Like the classification of finite simple groups. Who is going to read 1,000 pages of proof? Or even the proof of the four-color problem. I think people have to make mathematics interesting.

I think mathematics won't die soon. It will be around for some time, because there are a lot of beautiful things to be done. And mathematics is individual. I don't believe you can do mathematics with a group. Basically, it's individual. So it's easy to carry on. Mathematics does not need much equipment. It's not like other sciences. They need more material support than mathematics. So our subject will last for some time. Now, human civilization, I don't know how long it will last. That's a very much bigger problem. But mathematics itself, we will get along for some time.

At the age of 86 Chern continues to do mathematics. In recent years he has been especially interested in Finsler geometry, which he disscussed in a *Notices* article two years ago ("Finsler Geometry Is Just Riemannian Geometry Without the Quadratic Restriction" [September 1996]: pp. 959–963).

Chern: Finsler geometry is much broader than Riemannian geometry and can be treated in an elegant way. It will be the subject of the basic course on differential geometry within the next ten years in many universities. I have no difficulty in mathematics, so when I do mathematics, I enjoy it. Therefore I'm always doing mathematics, because the other things I cannot do. Like now, I am retired for many years, and people ask me if I still do mathematics. And I think my answer is, it's the only thing I can do. There is nothing else I can do. And this has been true throughout my life.

A Mathematician Needs Rich Imagination and Strong Resilience*: A Conversation at Nankai University in 2000

In 1985, the world-renowned mathematician Shiing-Shen Chern boldly envisioned China becoming a "mathematically advanced country" in the 21st century. This ambitious prediction, often referred to as the "Shiing-Shen Chern Conjecture," has inspired generations of Chinese mathematicians. Now, in the year 2000, widely considered the dawn of the 21st century, how close are we to realizing this dream?

At the start of the new millennium, I had the privilege of visiting Professor Chern at Nankai University, where he was spending the winter. Tragically, Chern had recently lost his beloved wife, Shih-Ning Cheng, with whom he shared over 60 years of companionship. She passed away peacefully, and both Chen and Nankai University decided that she would be laid to rest on the university's campus.

Chern's contributions to mathematics earned him prestigious accolades: He was among the first academicians of the Academia Sinica in 1948, a member of the US National Academy of Sciences since 1961, and a foreign academician of the Chinese Academy of Sciences in 1995. After a lifetime of global recognition, Chern returned to his roots, leaving an indelible legacy to China and to Nankai University.

---

* This article was first published in *Science* 52, no. 4 (2000). The original title was "Returning Home, Placing Hope in Nankai: An Interview with Shiing-Shen Chern." The interviewers were Zhang Dianzhou, Wang Shanping, and Ni Ming.

On January 28, 2000, I visited Professor Chern at his modest residence, located across from Nankai Yiyuan. With a bouquet of simple, elegant flowers in hand, I entered the home.

Despite his advanced age and recent loss, Chern greeted me with energy and warmth. His memory was as sharp as ever, his conversation lively and engaging. Over two sessions spanning six hours, he shared his reflections, thoughts, and wisdom without a hint of fatigue.

My primary purpose of this visit was to consult Professor Chern about details related to the upcoming publication of the Collected Works of Shiing-Shen Chern, being prepared by East China Normal University Press. Among other items, a *Shiing-Shen Chern Chronology* required careful fact-checking. As our conversation unfolded, it ventured far beyond mere details, touching on his life's journey and the future of mathematics in China. Below is an excerpt from our conversation, preserving the wisdom and insights of this great mind.

Q: Professor Chern, I understand you've recently written an article on "Mathematics and the Nobel Prize." Could you share your thoughts on this topic?

Chern: Yes, there is no Nobel Prize for mathematics. Originally, there was no prize for economics either; it was later added. Back then, if Swedish mathematicians had pushed harder, mathematics might have been included too. But mathematicians generally shy away from such activities. As a result, mathematics, a discipline closely intertwined with science, was excluded, while economics, which historically belonged more to the humanities, was added instead. That said, I think the absence of a Nobel Prize for mathematics might actually be a blessing in disguise. Mathematicians don't pursue their work for the sake of awards. They embrace a life of simplicity, far removed from utilitarian ambitions, dedicating themselves to pure research and the intrinsic beauty of mathematics.

Q: Even so, prestigious international prizes often serve as powerful motivators. You've won the Wolf Prize, and there's also the Fields Medal. These awards are still considered the highest honors in the mathematical community.

Chern: The early recipients of the Fields Medal were towering figures, and many of them made groundbreaking contributions to mathematics. However, I'm beginning to wonder whether the Fields Medal can maintain its high standard. I believe part of the problem lies in the way the Fields Medal is awarded. The International Mathematical Union doesn't have much funding, and the selection committee members can't afford to meet in person. Instead, they rely on correspondence to exchange views and cast their votes. As a result, the committee chair wields significant influence, which can lead to certain biases in the selection process. The Fields Medal also has an age limit, which was set when the award was not yet a major international prize back then. In hindsight, this age restriction now seems unreasonable.

Q: Nowadays, evaluations for qualifications and awards in the country involve a lot of "campaigning." This can lead to unfairness, can't it?

Chern: Whenever something is decided by a human vote, campaigning is inevitably involved. Some campaigning is necessary, like building relationships, participating in international activities, and collaborating on research to enhance mutual understanding. If you don't campaign, people won't even know what you're working on or who you are—how can they nominate you then? However, there are indeed some unhealthy forms of campaigning that can be quite annoying. Outsiders only see the awards and titles someone has and pay little attention to the substance. It's easier for them to manage that way. However, absolute fairness is very difficult to achieve.

Q: Does this mean that mathematicians who aren't good at campaigning can't gain high recognition?

Chern: No! As long as the mathematical work is truly outstanding, even if someone isn't an "academician" or hasn't won awards, they will still be respected. For example, from October 9 to 13, 2000, Nankai will host a workshop on the mathematical contributions of Wei-Liang Chow and Chen Guocai. Both cared little for fame or fortune. They didn't have titles like "academician" or major awards, but their work has proven to be increasingly important. Chow made significant contributions to algebraic geometry. Terms named after him appear five times in *Iwanami Mathematical Dictionary*, which is quite rare. Chen Guocai taught at several universities in the US without a high-profile position. He passed away quietly in Illinois in 1991. He was deeply creative, pursuing his own research without concern for others' understanding of his work. His contributions resemble the renowned de Rham Theorem. While de Rham Theorem connects differential forms with cohomology theory, Chen linked them with homotopy theory—a remarkably creative approach. Now people recognize the significance of his work.

Q: Mathematicians like Liao Shantao and Yan Zhida also seem to belong to this type. They work quietly, avoiding the spotlight, yet their achievements in mathematics are outstanding.

Chern: Liao earned his PhD under my supervision at the University of Chicago. He was very diligent—he'd draw the curtains during the day and lie in bed thinking about mathematics, ignoring everything else. At the time, McCarthyism was rampant in the US, but he was entirely oblivious to it. His English wasn't strong, and he didn't know a second foreign language, so he couldn't graduate for many years. At one point, I asked the faculty council to make a special exception for him, and they voted to approve it. He finally earned his PhD. Later, he returned to China and made significant contributions to dynamical systems. Yan could discuss problems with me when he was just a third-year undergraduate. He was very talented, loved reading Tang poetry, and would think about mathematics whenever he had spare time. Chow, Chen, Liao, and Yan didn't focus on campaigning but had distinctive personalities and unique insights. In fact, they were successful—far better than those who are all name and no substance.

Q: Your students and friends seem less inclined to campaign. Are you the same way?

Chern: No! I enjoy campaigning. I have many friends and like to blend work with life. I enjoy talking, dining, and collaborating with scientists of all ages, personalities, and backgrounds—primarily mathematicians. However, I dislike purely social events and avoid administrative roles, with the exception of being the director of mathematics institutes (both the Mathematical Sciences Research Institute and the NKIM, the latter after retirement).

Q: We've noticed that you've collaborated with many renowned mathematicians, achieving outstanding results. How do these collaborations usually come about?

Chern: The circumstances vary. For instance, Moser, a former president of the International Mathematical Union, once wrote to Wu Hung-Hsi asking about a problem in Elie Cartan's works. Wu relayed the question to me. Since I was familiar with Cartan's work, we started collaborating, and eventually produced a highly influential paper published in *Acta Mathematica*. I've also collaborated extensively with Phillip Griffiths. While he was at Princeton, he read my mimeographed book *Complex Manifolds* and found it very interesting. Each summer, he came to Berkeley to discuss problems with me—not just as a courtesy visit. As we had more mathematical discussions, we began collaborating. Later, he joined Berkeley's faculty and became a full professor. Over time, he moved to Princeton, then Harvard, followed by Duke University, where he held a senior administrative position. Eventually, he became the director of the Institute for Advanced Study at Princeton. Griffiths was also the secretary of the International Mathematical Union and may eventually become its president. Collaboration grows naturally through frequent interactions and discussions. For example, my collaborations with Friedrich Hirzebruch, Raoul Bott, and other renowned mathematicians also arose through regular interactions.

Q: Recently, the term "Chern-Simons-Witten invariant" has been frequently mentioned in academic literature. Can you tell us more about Simons?

Chern: Simons is quite the legendary figure. He graduated from MIT and had a strong interest in differential geometry, which led him to Berkeley to study with me in 1959. At that time, I was in Europe, so he taught himself. Once he grasped the material, he posted notices inviting others to attend his lectures—and many did, including professors! When I returned to Berkeley, Simons already had a mentor, but we continued to interact frequently. He was extraordinarily talented, balancing his studies with a budding career in business. Despite his relatively limited academic experience, he was appointed chair of the mathematics department at the State University of New York at Stony Brook. When we collaborated on the invariants, we had no idea they would eventually have applications in physics. Years later, Edward Witten, the 1990 Fields Medalist, applied them to his research in physics—something we could not have foreseen. This exemplifies how great mathematical discoveries often start without a clear application but later prove to be useful. For this reason, I oppose the rigid division between pure and applied mathematics. Simons eventually transitioned into currency trading and achieved remarkable financial success. Mathematicians, like everyone else, follow diverse paths, and I enjoy connecting with them all.

Q: One of your students, who won a lottery jackpot, donated $1 million to establish the "Chern Lectures." What's the story behind that?

Chern: That student is Robert Uomini, a Berkeley graduate. As an undergraduate, his academic performance was average, and he sought my help to get into graduate school. I thought he had potential, so I wrote him a recommendation letter. He eventually earned his PhD and went on to work at a computer company. Uomini had a habit of buying lottery tickets, and one day, he won $22 million. To express his gratitude, he donated $1 million to establish the "Chern Lectures" at Berkeley. The endowment's interest is used to invite a world-class mathematician each year to give lectures. So far, we've welcomed exceptional mathematicians like Sir Michael Atiyah, Richard Stanley, Friedrich Hirzebruch, Jean-Pierre Serre, Yuri Manin, and in 1999, Michael Artin (the son of the renowned mathematician Emil Artin). All of them are among the very best in the field.

Q: Scientific research in China now emphasizes "innovation" heavily. What's your perspective on innovation in mathematics?

Chern: Mathematics is a discipline where there is only one winner—there's no second place. Whether in China or abroad, if a result has already been published, it cannot be published again. In this sense, all mathematical research is inherently innovative. In the past, many university presidents in China were mathematicians. I believe this is because mathematical results are recognized for their innovation, making it easier to gain recognition. That said, there's a clear distinction between good mathematics and less impactful work. Unfortunately, many discussions about "innovation" remain superficial and slogan-driven.

Q: How would you define "good mathematics"?

Chern: That's a difficult question, but mathematicians generally have a shared understanding of what constitutes "good mathematics." For instance, consider Fermat's Last Theorem: $x^n + y^n = z^n$, where $n \geq 3$, has no integer solutions for $xyz \neq 0$. The statement is simple enough to explain to anyone, but its proof is immensely challenging and profoundly deep. Another example is the study of equations, which has evolved over millennia—from linear equations to quadratic equations, systems of equations, differential equations, integral equations, and beyond. This progression is unending and exemplifies the enduring vitality of mathematics. On the other hand, "mediocre mathematics" often focuses on trivial or irrelevant problems. Some work lacks originality, merely refining existing results with minor technical improvements. While such efforts may serve as valuable practice for beginners, they shouldn't be considered the pinnacle of mathematical achievement.

Q: In 1991, Professor Zhang Dianzhou visited you in Berkeley, and you mentioned the terms "mathematical craftsman" and "mathematical architect." Do you still hold this view?

Chern: Absolutely. Mathematical research requires two key abilities: imagination and resilience to tackle problems. Imagination allows one to propose theoretical frameworks, construct concepts, pose problems, and identify the essential points. Resilience to tackle problems, on the other hand, is needed to construct specific objects, discover invariants, and compute quantities accurately. It's akin to building a skyscraper—you need architects to design and craftsmen to construct. In mathematics, we need both "mathematical architects" and "mathematical craftsmen." Both roles are indispensable, and the greatest mathematicians excel in both—they design and build their own creations.

Q: In recent years, you've been advocating for Finsler geometry. Do you believe it represents "good mathematics"?

Chern: I believe it does. Transitioning from Riemannian geometry to Finsler geometry is a natural progression. In fact, Finsler geometry generalizes the ideas Riemann originally proposed. It's also connected to the last of Hilbert's famous 23 problems—a variation problem. I've uncovered relationships in Finsler geometry that weren't previously recognized, fundamentally transforming the field. A new book on this topic is about to be published. Sadly, despite advocating for Finsler geometry in China for five years, there has been little uptake.

Q: I heard it is because it is "lacking a strong background" and "not a hot topic"?

Chern: Riemann-Finsler geometry originates from Riemann's groundbreaking 1854 paper. At the time, the focus of mathematics was on analysis, so it was overlooked. It later gained prominence through its applications in general relativity. Riemann only discussed the special case of quadratic metrics, now called Riemannian geometry, which is relatively simple but incredibly profound. Today, we know that the general case can be handled just as well. Please read our new book. Finsler geometry will undoubtedly find applications, such as in solid-state physics. The criticisms you mentioned only highlight the critics' ignorance, so they don't surprise me.

Q: The current president of the International Mathematical Union, Jacob Palis, is from Brazil. Some say Brazil, India, and China are the three largest developing countries. You've had extensive interactions with the mathematical communities in all three. Which country do you think has the strongest mathematics?

Chern: China. The reason is simple: China has a deep tradition of scholarship. In Brazil, for example, it's challenging to convince the general public of the importance of mathematics.

Q: So, you remain confident that China will become a "mathematically advanced country" in the 21st century?

Chern: Absolutely. The mathematical talent of the Chinese people is beyond question. What's needed now is sustained effort. Mathematics can develop independently and doesn't require extensive resources. Compared to other sciences, it's relatively easy to advance, though support is still essential.

Mathematics Simplifies Science*: A Conversation with *Guangming Daily* Reporter Cai Chuang

Cai: You have dedicated your life to mathematics. For many, mathematics seems like an incredibly profound and abstract discipline, and learning it feels daunting.

Chern: Learning anything fundamentally depends on one's own effort and initiative. Some people finish their studies, toss their textbooks aside, and hope to never revisit them—naturally, they won't learn much that way. To truly master a subject, it's not enough to go through it once. As the old Chinese saying goes, "Read a book a hundred times, and its meaning will reveal itself."

I've always been able to connect problems from various contexts and see the relationships between different pieces of knowledge. Personally, I don't find mathematics difficult. If someone feels it's insurmountable from the start, perhaps mathematics isn't the right path for them. After all, we can't expect everyone to become scholars. But even for ordinary individuals, learning some mathematics—whether a little or a lot—is essential. Mathematics trains logical reasoning, which is invaluable. A truly significant book in this regard is *Euclid's Elements*. It's one of the best-selling books in the West, second only to the Bible. Many parents used it to train their children's

* First published in *Guangming Daily*, January 31, 2002, B01.

reasoning abilities. This book was brought to China by Matteo Ricci and translated into Chinese by Xu Guangqi. Unfortunately, in our education system, mathematics has become a required exam subject because so many other fields depend on it.

However, I believe academic success fundamentally depends on the individual. You must be self-directed, know what you want to achieve, and continually strive to improve. Talent in mathematics can't be cultivated by someone else—it must come from within. If you always need someone to tell you what to do or how to do it, you won't go far.

To excel in your work, you must not only master your discipline but also acquire knowledge from other fields. Don't wait for instructions from your teacher. You should know what needs to be done and how to do it, and strive to innovate on your own.

Cai: You are currently teaching foundational mathematics courses at Nankai University. For many undergraduates, this is an extraordinary opportunity.

Chern: People should not be ranked as high or low, and knowledge should not be divided into shallow or profound. Undergraduate courses form the cornerstone of learning, and I'll always find foundational material fascinating. Even the most advanced research is built on these basics. Mathematics is deep and layered—you often need to pass through one stage to reach the next. But once you've progressed, you'll see the discipline as a unified whole. Teaching the fundamentals of mathematics is deeply rewarding for me. I present the material in my own way, incorporating insights from my research, and the students grasp it well. In my view, there is no real distinction between pure and applied mathematics. Truly foundational concepts often find practical applications, and applied work frequently becomes the basis for future foundational developments—they ultimately converge. Dividing pure and applied mathematics is a matter of convenience because the field is vast, and any individual can only focus on a particular aspect at a given time. But in reality, mathematics is an interconnected whole.

Cai: Historically, mathematics hasn't been considered a particularly developed discipline in China. When you first decided to study mathematics, would you say you were taking an unconventional path?

Chern: I'm from Jiaxing, Zhejiang. My father had a job in Tianjin, so I moved there at a very young age. I spent over a decade studying in Tianjin before attending the Tsinghua University Institute for Advanced Study for four years. Later, I traveled to Germany to further my studies. In 1937, I returned to China from Europe, intending to work at Tsinghua University. However, due to the War of Resistance Against Japanese Aggression, I ended up at the Temporary University in Changsha, a collaborative effort by Peking University, Tsinghua University, and Nankai University. Eventually, I moved to the National Southwest Associated University in Kunming, which was by then 1938.

Historically, China has placed great emphasis on interpersonal relationships and paid much attention to social sciences and societal structures. In contrast, Western nations have focused on the development of natural sciences and the exploration of the natural world. I believe ancient Chinese mathematics leaned heavily toward practical applications, which can be seen as a limitation of ancient Chinese science. When I was a student, I came across articles written by Japanese scholars claiming that the Chinese were only capable of studying history, not science. This assertion was utterly absurd. The French writer Romain Rolland once wrote a book recounting the experiences of German musicians in Rome during the Middle Ages. At that time, the Romans mocked the Germans, saying, how can these barbarians possibly understand music? Yet not long after, Germany gave the world Bach and Beethoven. When I was in Germany, my academic performance was excellent. Today, Chinese students perform exceptionally well in the International Mathematical Olympiad. I hope some of these bright students will pursue careers in mathematics—it's a field full of potential and fascination. Among the younger generation, many have already reached international standards. For example, Professor Shing-tung Yau has won the Fields Medal at the International Congress of Mathematicians, and others, such as Xiao Yintang, Mo Yiming, Tian Gang, Xiang Wuyi, and Li

Weiguang, have emerged as outstanding mathematicians. I hope that one day China will become a mathematically advanced country.

Cai: After decades of exploring the challenging field of mathematics, how would you define mathematics?

Chern: In its simplest form, mathematics began with addition, subtraction, multiplication, and division—basic arithmetic. But as time passed, these rudimentary tools were no longer sufficient. Newton's monumental contribution was transforming motion into mathematics. Motion, while incredibly complex, follows certain laws, and these laws are described by mathematics. Mathematics simplifies science.

The scope of mathematics continues to expand. Mathematicians must constantly broaden their understanding and explore new avenues for growth. The mathematics we study today is new, but it is deeply rooted in the old starting point, which is foundational mathematics. Mathematics is always evolving, not only in its applications but also in its theoretical foundations. Mathematicians must keep pushing forward, always exploring and discovering.

Research is not simply a matter of assigning a problem to solve in exchange for funding. The best research often starts with uncertainty—you don't know what the problem is, where the solution lies, or when it will be solved. True research requires an atmosphere of freedom, where everyone can explore independently. Mathematics is an ongoing process; it is always evolving.

Popular Mathematics Lecture

From Triangles to Manifolds*

This article offers an engaging and accessible overview of the development of global differential geometry. It explores how tools from topology have driven progress in areas such as partial differential equations, global analysis, unified field theories in particle physics, and DNA theory in molecular biology. The author highlights the interplay between local and global topological properties, emphasizing that "the Euler characteristic serves as a source of global invariants." Additionally, the author notes that "most known global structures on manifolds are closely tied to even dimensions" and expresses hope that more attention will be given to the study of odd-dimensional manifolds in the future.

Geometry

I believe I am expected to tell you all about geometry; what it is, its developments through the centuries, its current issues and problems, and, if possible, a peep into the future. The first question does not have a clear-cut answer. The meaning of the word geometry changes with time and with the speaker. With Euclid, geometry consists of the logical conclusions drawn from a set of axioms. This is clearly not sufficient with the horizons of geometry ever widening. Thus in 1932 the great geometers O. Veblen and J. H. C. Whitehead said, "A branch of mathematics is called geometry, because the name seems good on emotional and traditional grounds to a sufficiently large number of competent people."[1] This opinion was enthusiastically seconded by the great French geometer Elie Cartan.[2] Being an analyst himself, the great

---

* This is the "Faculty Research Lecture" delivered by Mr. Shiing-Shen Chern on April 7, 1978, at the University of California, Berkeley. Similar lectures were later given in Beijing, Changchun, and other locations. The lecture was translated into Chinese by You Chengye and published in the *Chinese Journal of Nature* 2, no. 8 (1979).

American mathematician George Birkhoff mentioned a "disturbing secret fear that geometry may ultimately turn out to be no more than the glittering intuitional trappings of analysis."[3] Recently my friend André Weil said: "The psychological aspects of true geometric intuition will perhaps never be cleared up. At one time it implied primarily the power of visualization in three-dimensional space. Now that higher-dimensional spaces have mostly driven out the more elementary problems, visualization can at best be partial or symbolic. Some degree of tactile imagination seems also to be involved."[4]

At this point it is perhaps better to let things stand and turn to some concrete topics.

Triangles

Among the simplest geometrical figures is the triangle, which has many beautiful properties. For example, it has one and only one inscribed circle and also one and only one circumscribed circle. At the beginning of this century the nine-point circle theorem was known to almost every educated mathematician. But its most intriguing property concerns the sum of its angles. Euclid says that it is equal to 180°, or π by radian measure, and deduces this from a sophisticated axiom, the so-called parallel axiom. Efforts to avoid this axiom failed. The result was the discovery of non-Euclidean geometries in which the sum of angles of a triangle is less or greater than π, depending on whether the geometry is hyperbolic or elliptic. The discovery of hyperbolic non-Euclidean geometry, in the eighteenth century by Gauss, John Bolyai, and Lobatchevsky, was one of the most brilliant chapters in human intellectual history.

The generalization of a triangle is an n-gon, a polygon with n sides. By cutting the n-gon into $n-2$ triangles, one sees that the sum of its angles is $(n-2)\pi$. It is better to measure the sum of the exterior angles! The latter is equal to 2π, for all n-gons, including triangles.

Curves in the Plane; Rotation Index and Regular Homotopy

By applying calculus we can consider smooth curves and closed smooth curves in the plane, i.e., curves with a tangent line everywhere and varying continuously. As a point moves along a closed smooth (oriented) curve C once, the lines through a fixed-point O and parallel to the tangent lines of C rotate through an angle $2n\pi$ or rotate n times about O. This integer n is called the rotation index of C (see fig. 1). A famous theorem in differential geometry says that if C is a simple curve, i.e., if C does not intersect itself, $n = \pm 1$.

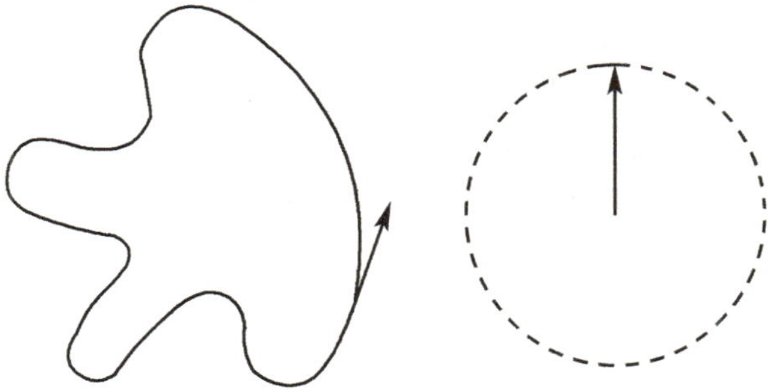

Fig. 1

Clearly, there should be a theorem combining the theorem on the sum of exterior angles of an n-gon and the rotation index theorem of a simple closed smooth curve. This is achieved by considering the wider class of simple closed sectionally smooth curves. The rotation index of such a curve can be defined in a natural way by turning the tangent at a corner an amount equal to the exterior angle (see fig. 2). Then the rotation index theorem above remains valid for simple closed sectionally smooth curves. In the particular case of an n-gon formed by straight segments, this reduces to the statement that the sum of its exterior angle is 2π.

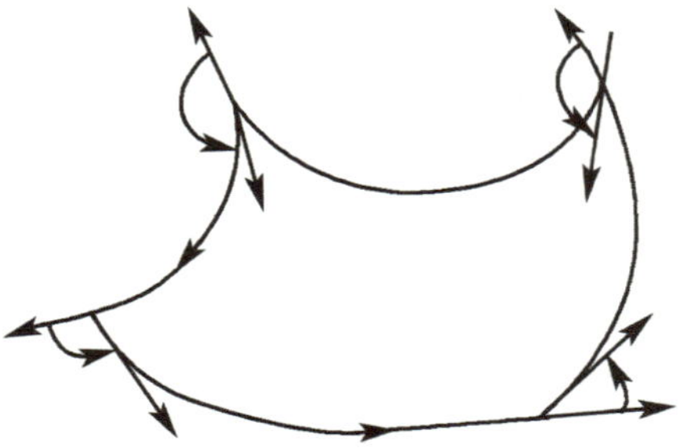

Fig. 2

This theorem can be further generalized. Instead of simple closed curves we can allow closed curves to intersect themselves. A generic self-intersection can be assigned a sign. Then, if the curve is properly oriented, the rotation index is equal to one plus the algebraic sum of the number of self-intersections (see fig. 3). For example, figure "8" has a rotation index of zero.

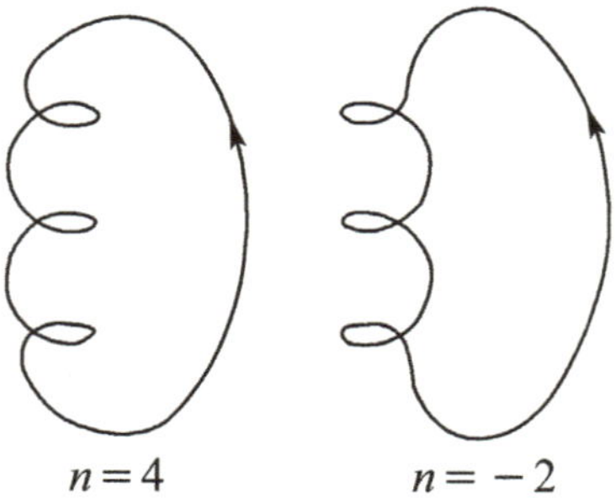

Fig. 3

A fundamental notion in geometry, or in mathematics in general, is deformation or homotopy. Two closed smooth curves are said to be regularly homotopic if one can be deformed to the other through a family of closed smooth curves. Since the rotation index is an integer and varies continuously in the family, it must remain constant, i.e., it keeps the same value when the curve is regularly deformed. A remarkable theorem of Graustein-Whitney says that the converse is true[5]: Two closed smooth curves with the same rotation index are regularly homotopic.

It is a standard practice in mathematics that in order to study closed smooth curves in the plane it is more profitable to look at all curves and to put them into classes, the regular homotopy classes in this case being an example. This may be one of the essential methodological differences between theoretical science and experimental science, where such a procedure is impractical. The Graustein-Whitney Theorem says that the only invariant of a regular homotopy class is the rotation index.

Euclidean Three-Space

From the plane we pass to the three-dimensional Euclidean space where the geometry is richer and has distinct features. Perhaps the nicest space curve which does not lie in a plane is a circular helix. It has constant curvature and constant torsion and is the only curve admitting ∞^1 rigid motions. There is an essential difference between right-handed and left-handed helices (see fig. 4), depending on the sign of the torsion; a right-handed helix cannot be congruent to a left-handed one, except by a mirror reflection. Helices play an important role in mechanics. From a geometrical viewpoint it may not be an entire coincidence that the Crick-Watson model of a DNA-molecule is double-helical. A double helix has interesting geometrical properties. In particular, by joining the end points of the helices by segments or arcs, we get two closed curves. In three-dimensional space they have a linking number L (see fig. 5).

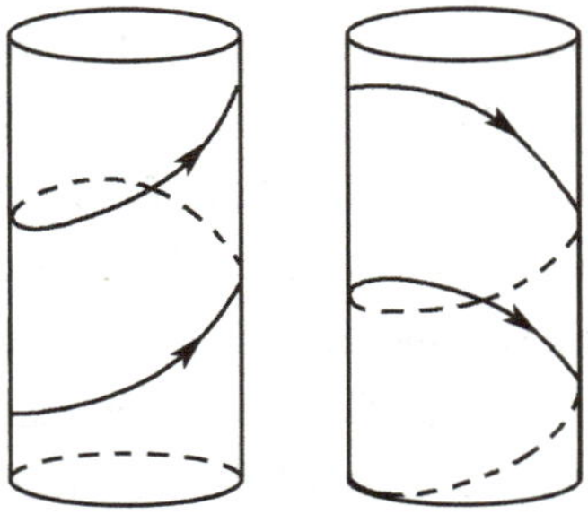

Fig. 4

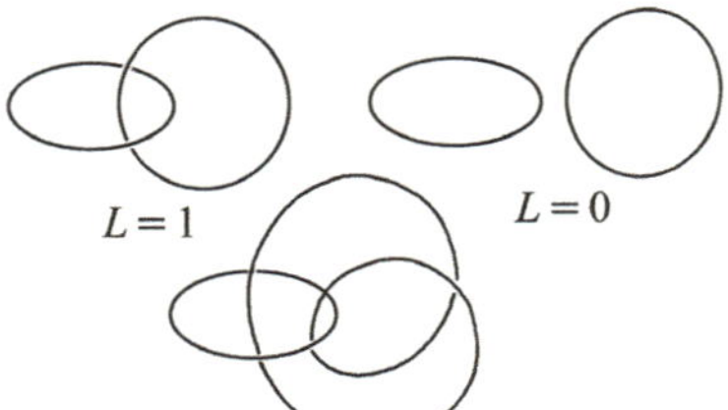

Fig. 5

A recent controversial issue in biochemistry, raised by the mathematicians William Pohl and George Roberts, is whether the chromosomal DNA is double-helical. In fact, if it is, it will have two closed strands with a linking number of the order of 300,000. The molecule is replicated by the separation of the strands and the formation of the complementary strand of each. With such a large linking number Pohl and Roberts showed that the replication process would have severe mathematical difficulties. Thus the double-helical structure of the DNA-molecule, at least for the chromosome, has been questioned.[6]

The linking number L is determined by the formula of James H. White[7]:

$$T + W = L \tag{1}$$

T is the total twist and W is the writhing number. The latter can be experimentally measured and changes by the action of an enzyme. This formula is of fundamental importance in molecular biology. Generally DNA molecules are long. In order to store them in limited space, the most economical way is to writhe and coil them. These discussions could indicate the beginning of a stochastic geometry, with the main examples drawn from biology.

In a three-dimensional space surfaces have far more important properties than curves. Gauss's fundamental work elevated differential geometry from a chapter of calculus to an independent discipline. His "Disquisitiones Generales Circa Superficies Curvas" (1827) is the birth certificate of differential geometry. The main idea is that a surface has an intrinsic geometry based on the measure of arc length alone. From the element of arc other geometric notions, such as the angle between curves and the area of a piece of surface, can be defined. Plane geometry is thus generalized to any surface Σ based only on the local properties of the element of arc. This localization of geometry is both original and revolutionary. In place of the straight lines are the geodesics, the "shortest" curves between any two points (sufficiently close). More generally, a curve on Σ has a "geodesic curvature," generalizing the curvature of a plane curve and geodesics are the curves whose geodesic curvature vanishes identically.

Let the surface Σ be smooth and oriented. At every point P of Σ there is a unit normal vector $v(P)$ which is perpendicular to the tangent plane to Σ at P (see fig. 6). The vector $v(P)$ can be viewed as a point of the unit sphere S_0 with center at the origin of the space. By sending P to $v(P)$ we get the Gauss mapping:

$$g: \Sigma \longrightarrow S_0 \tag{2}$$

The ratio of the element of the area of S_0 by the element of the area of Σ under this mapping is called the Gaussian curvature. Gauss's "remarkable theorem" says that the Gaussian curvature depends only on the intrinsic geometry of Σ. In fact, in a sense it characterizes this geometry. Clearly the Gaussian curvature is zero if Σ is the plane.

Fig. 6

As in plane geometry we consider on Σ a domain D bounded by one or more sectionally smooth curves. D has an important topological invariant $\chi(D)$, called its Euler characteristic, which is most easily defined as follows: By appropriately subdividing D into polygons and letting v, e, and f represent the number of vertices, edges, and faces, respectively, we have

$$\chi(D) = \mathrm{v} - \mathrm{e} + \mathrm{f} \tag{3}$$

(Euler Polyhedral Theorem was known before Euler, but Euler seems to have been the first one to recognize explicitly the importance of the "alternating sum.")

The Gauss-Bonnet formula in surface theory is

$$\sum \text{ext angles} + \int_{\partial D} \text{geod curv} + \iint_{D} \text{Gaussian curv} = 2\pi\chi(D) \qquad (4)$$

where ∂D is the boundary of D. If D is a plane domain, the Gaussian curvature is 0. If in addition the domain is simply connected, we have $\chi(D) = 1$. Then this formula reduces to the rotation index theorem discussed in section 3. We are indeed a long way from the sum of angles of a triangle.

Generalizing the geometry of closed plane curves we can consider closed oriented surfaces in space. The generalization of the rotation index is the degree d of the Gauss mapping g in (2). The precise definition of the d is sophisticated. Intuitively it is the number of times that the image $g(\Sigma)$ covers S_0, counted with sign. Unlike the plane, where the rotation index can be any integer, the degree d is completely determined by the topology of Σ; it is equal to

$$d = \frac{1}{2}\chi(\Sigma) \qquad (5)$$

For the embedded unit sphere this degree is +1 independently of its orientation. A surprising result by S. Smale[8] says that the two oppositely oriented unit spheres are indeed regularly homotopic or, more intuitively, that the unit sphere can be turned inside out through a regular homotopy. It is essential that at each stage of the homotopy the surface has a tangent plane everywhere, but is allowed to intersect itself.

From Coordinate Spaces to Manifolds

It was Descartes who in the seventeenth century revolutionized geometry by using coordinates. Quoting Hermann Weyl, "The introduction of numbers as coordinates was an act of violence."[9] From now on, paraphrasing Weyl, figure and number, like angel and devil, fight for the soul of every geometer. In the plane, the Cartesian coordinates (x,y) of a point represent its signed distances to two fixed mutually perpendicular lines (the coordinate axes). A straight line is the locus of points satisfying the linear equation:

$$ax + by + c = 0 \qquad (6)$$

The result is the translation of geometry into algebra.

Once the door was opened for analytic geometry, other coordinate systems came into play. Among them are polar coordinates in the plane and spherical coordinates, cylindrical coordinates in space, and elliptic coordinates in the plane and in space. The latter are adapted to the confocal quadrics and are particularly suited to the study of the ellipsoids, which include our earth.

There is also a need for higher dimensions. For even if we start with a three-dimensional space, the theory of relativity calls for the inclusion of time as a fourth dimension. On a more elementary level, recording the motion of a particle, including its velocity, requires six coordinates (the hodograph). All the continuous functions in one variable form an infinite-dimensional space. Those that are square-integrable form a Hilbert space, which can be coordinatized by an infinite sequence of coordinates. Such a viewpoint, of considering all functions with prescribed properties, is fundamental in mathematics.

From the proliferation of coordinate systems, it is natural to have a theory of coordinates. General coordinates need only the property that they can be identified with points; i.e., there is a one-to-one correspondence between points and their coordinates-their origin and meaning are inessential.

If you find it difficult to accept general coordinates, you will be in good company. It took Einstein seven years to pass from his special relativity in 1908 to his general relativity in 1915. He explained the long delay in the following words: "Why were another seven years required for the construction of the general theory of relativity? The main reason lies in the fact that it is not so easy to free oneself from the idea that coordinates must have an immediate metrical meaning."[10]

After being served by coordinates in the study of geometry, we now wish to be free from their bond. This leads to the fundamental notion of a manifold. A manifold is described locally by coordinates, but the latter are subject to arbitrary transformations. In other words, it is a space with transient or relative coordinates (principle of relativity). I would compare the concept with the introduction of clothing to human life. It was a historical event of the utmost importance that human beings began to clothe themselves. No less significant was the ability of human beings to change their clothing. If geometry is the human body and coordinates are clothing, then the evolution of geometry has the following comparison.

Synthetic geometry	Naked man
Coordinate geometry	Primitive man
Manifolds	Modern man

A manifold is a sophisticated concept even for mathematicians. For example, a great mathematician such as Jacques Hadamard "felt insuperable difficulty … in maintaining more than a rather elementary and superficial knowledge of the theory of Lie groups,"[11] a notion based on that of a manifold.

Manifolds; Local Tools

With coordinates practically meaningless there is a need for new tools in studying manifolds. The key word is invariance. Invariants are of two kinds: local and global. The former refers to the behavior under a change of the local coordinates, while the latter is the global invariant of the manifold, an example being the topological invariant. Two of the most important local tools are the exterior differential calculus and Ricci's tensor analysis.

An exterior differential form is the integrand of a multiple integral, such as

$$\iint_D P\mathrm{d}y\mathrm{d}z + Q\mathrm{d}z\mathrm{d}x + R\mathrm{d}x\mathrm{d}y \tag{7}$$

in (x,y,z)-space, where P, Q, and R are functions in (x,y,z), and D is a two-dimensional domain. It is observed that a change of variables in D (supposed to be oriented) will be taken care of automatically if the multiplication of differentials is anti-symmetric, i.e.,

$$\mathrm{d}y \wedge \mathrm{d}z = -\,\mathrm{d}z \wedge \mathrm{d}y \tag{8}$$

where the symbol $\wedge$ is used to denote exterior multiplication. It is also more suggestive to introduce the exterior two-form:

$$w = P\mathrm{d}y \wedge \mathrm{d}z + Q\mathrm{d}z \wedge \mathrm{d}x + R\mathrm{d}x \wedge \mathrm{d}y \tag{9}$$

and to write the integral (7) as a pairing (D, w) of the domain D and the form w.

For if the same is done in n-space, then Stokes Theorem can be written as

$$(D, \mathrm{d}w) = (\partial D, w) \tag{10}$$

where D is an r-dimensional domain and w is an exterior $(r\text{–}l)$-form; ∂D is the boundary of D and dw is the exterior derivative of w and is an r-form. Formula (10), the fundamental formula in multi-variable calculus, shows that ∂ and d are adjoint operators. The remarkable fact is that, while the boundary operator ∂ on domains is global, the exterior differentiation operator d on forms is local. This makes d a powerful tool. When applied to a function (= 0-form) and a 1-form, it gives the gradient and the curl, respectively. All the smooth forms, of all degrees (≤ dim of manifold), of a differentiable manifold constitute a ring with the exterior differentiation operator d. Elie Cartan used exterior differential calculus most efficiently in local problems of differential geometry and partial differential equations. The global theory was founded by Georges de Rham, after initial work of Poincaré. This will be discussed in the next section.

In spite of its importance the exterior differential calculus is inadequate in describing the geometrical and analytical phenomena on a manifold. A broader concept is Ricci's tensor analysis. Tensors are based on the fact that a manifold, being smooth, can be approximated at every point by a linear space, called its tangent space. The tangent space at a point leads to associated tensor spaces. Differentiation of tensor fields needs an additional structure, called an affine connection. If the manifold has a Riemannian or Lorentzian structure, the corresponding Levi-Civita connection will serve the purpose.

Homology

Historically a systematic study of the global invariants of a manifold began with combinatorial topology. The idea is to decompose the manifold into cells and see how they fit together. (The decomposition satisfies some mild conditions, which we will not specify.) In particular, if M is a closed manifold of dimension n and a_k denotes the number of k cells of the decomposition, $k = 0, 1, \ldots, n$, then, as a generalization of (3), the Euler-Poincaré characteristic of M is defined by

$$\chi(M) = \alpha_0 - \alpha_1 + \cdots + (-1)^n \alpha_n \tag{11}$$

The basic notion in homology theory is that of a boundary. A chain is a sum of cells with multiplicities. It is called a closed chain if it has no boundary, i.e., if its boundary is 0. The boundary of a chain is a closed chain (see fig. 7). The number of linearly independent k-dimensional closed chains modulo k-dimensional boundaries is a finite integer, denoted b_k, called the k-th Betti number. The Euler-Poincaré formula says

$$\chi(M) = b_0 - b_1 + \cdots + (-1)^n b_n \tag{12}$$

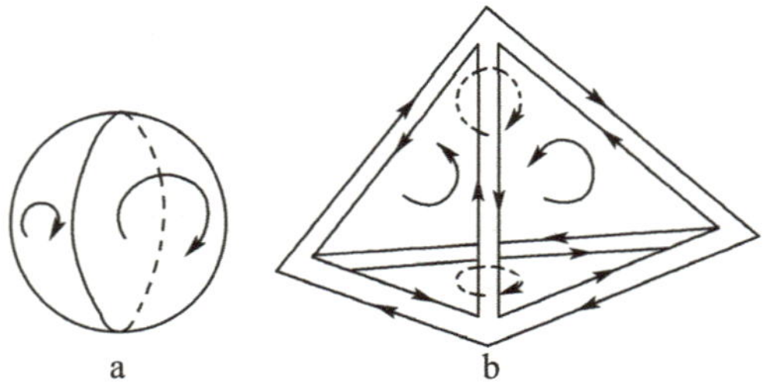

Fig. 7

The Betti numbers b_k, and hence $\chi(M)$ itself, are topological invariants of M. In other words, they are independent of the decomposition and remain invariant under a topological transformation of M. This and more general statements could be considered the fundamental theorems of combinatorial topology. After the path-breaking works of Poincaré and L. E. J. Brouwer, combinatorial topology blossomed in the US in the 1920s under the leadership of Veblen, Alexander, and Lefschetz.

While this is an effective way in deriving topological invariants, the danger in cutting a manifold is that it might be "killed." Precisely, this means that by using a combinatorial approach we may lose sight of the relations of the topological invariants with local geometrical properties. It turns out that, while homology theory depends on the boundary operator ∂, there is a dual cohomology theory based on the exterior differentiation operator d, the latter being a local operator.

The resulting de Rham cohomology can be summarized as follows: The operator d has the fundamental property that, when applied repeatedly it gives the 0 form; that is, for any k-form α, the exterior derivative of the $(k + 1)$-form $d\alpha$ is 0. This corresponds to the geometrical fact that the boundary of any chain (or domain) has no boundary (see equation [10]). A form a is called closed, if $d\alpha = 0$.

It is called a derived form, if there exists a form β, of degree $(k - 1)$, such that it can be written $\alpha = \mathrm{d}\beta$. Thus α derived form is always closed. Two closed forms are called cohomologous if they differ by a derived form. All the closed k-forms which are cohomologous to each other constitute the k-dimensional cohomology class. The remarkable fact is that, while the families of k-forms, closed k-forms, derived k-forms are immensely large, the k-dimensional cohomology classes constitute a finite-dimensional linear space whose dimension is the k-th Betti number b_k.

De Rham cohomology is the forerunner of sheaf cohomology, which was founded by J. Leray[12] and perfected and applied with great success by H. Cartan and J. P. Serre.

Vector Fields and Generalizations

On a manifold M it is natural to consider continuous vector fields, i.e., the attachment of a tangent vector to each point, varying in a continuous manner. If the Euler-Poincaré characteristic $\chi(M) \neq 0$, there is at least one point of M at which the vector vanishes. In other words, when the wind blows there is at least one spot on earth with no wind (for the Euler characteristic of the two-dimensional sphere is equal to 2). More precisely, at an isolated zero of a continuous vector field, an integer, called the index, can be defined, which describes to a certain extent the behavior of the vector field at the zero, i.e., whether it is a source, a sink, or otherwise. No matter what the vector field is, so long as it is continuous and has only a finite number of zeros, then the Poincaré-Hopf Theorem says that the sum of its indices at all the zeros is a topological invariant which is precisely $\chi(M)$.

This is a statement on the tangent bundle of M, i.e., the collection of the tangent spaces of M. More generally, a family of vector spaces parametrized by a manifold M and satisfying a local product condition is called a vector bundle over M.

A fundamental question is whether such a bundle is globally a product. The above discussion shows that the tangent bundle is not a product if $\chi(M) \neq 0$; for if it were a product, there would exist a continuous vector field which is nowhere 0. The existence of a space which is locally but not globally a product, such as the tangent bundle of a manifold M with $\chi(M) \neq 0$, is not easy to visualize; geometry thus enters a more sophisticated phase.

To describe the global deviation of a vector bundle from a product space the first invariants are the so-called characteristic cohomology classes. The Euler-Poincaré characteristic is the simplest of the characteristic classes.

The Gauss-Bonnet formula (4) in section 4 takes the particularly simple form when the surface Σ has no boundary:

$$\iint K\mathrm{d}A = 2\pi\chi(M) \tag{4a}$$

In this formula, K is the Gaussian curvature and $\mathrm{d}A$ is the element of area. Formula (4a) is of paramount importance because it expresses the global invariant $\chi(\Sigma)$ as the integral of a local invariant, which is perhaps the most desirable relationship between local and global properties. This result has a wide generalization. Let

$$\pi: E \longrightarrow M \tag{13}$$

be a vector bundle. The generalization of a tangent vector field on M is a section of the bundle, i.e., a smooth mapping $s: M \longrightarrow E$, such that the composition $\pi^\circ s$ is the identity. Since E is only locally a product, the differentiation of s needs an additional structure, usually called a connection. The resulting differentiation, called covariant differentiation, is generally not commutative. The notion of curvature is a measure of the noncommutativity of covariant differentiation. Suitable combinations of the curvature give rise to differential forms that represent characteristic cohomology classes in the sense of the de Rham cohomology, of which the Gauss-Bonnet formula (4a) is the simplest example.[13] I believe that the concepts of vector bundles, connections, and curvature are so fundamental and so simple that they should be included in any introductory course on multivariable calculus.

Elliptic Differential Equations

When an n-dimensional manifold M is equipped with a Riemannian metric, there exists an operator $*$ that transforms a k-form α into an $(n-k)$-form $*\alpha$. This corresponds to taking the orthogonal complement of a linear subspace in the tangent space. Using the operator $*$ and the exterior derivative d, we introduce the codifferential:

$$\delta = (-1)^{nk+n+1} * d * \tag{14}$$

and the Laplacian

$$\Delta = \mathrm{d}\delta + \delta\mathrm{d} \tag{15}$$

Then the operator δ sends a k-form to a $(k-1)$-form and Δ sends a k-form to a k-form. A form α satisfying

$$\Delta\alpha = 0 \tag{16}$$

is called harmonic. A harmonic form of degree 0 is a harmonic function in the usual sense.

The equation (16) is an elliptic partial differential equation of the second order. If M is closed, all its solutions form a finite-dimensional vector space. By a classical theorem of Hodge this dimension is exactly the k-th Betti number b_k. It follows by (12) that the Euler characteristic can be written:

$$\chi(\mathrm{M}) = d_e - d_o \tag{17}$$

where d_e (respectively, d_o) is the dimension of the space of harmonic forms of even (respectively odd) degree. The exterior derivative d is itself an elliptic operator and (17) can be regarded as expressing $\chi(M)$ as the index of an elliptic operator. The latter is, for any linear elliptic operator, equal to the dimension of the space of solutions minus the dimension of the space of solutions of the adjoint operator.

The expression of the index of an elliptic operator as the integral of a local invariant culminates in the Atiyah-Singer Index Theorem. It includes as special cases many famous theorems, such as the Hodge Signature Theorem, the Hirzebruch Signature Theorem, and the Riemann-Roch Theorem for complex manifolds. An important by-product of this study is the recognition of the need to consider pseudo-differential operators on manifolds, which are more general than differential operators.

Elliptic differential equations and systems are closely enmeshed with geometry. The Cauchy-Riemann differential equations, in one or more complex variables, are at the foundation of complex geometry. Minimal varieties are solutions of the Euler-Lagrange equations of the variational problem minimizing the area. These equations are quasi-linear. Perhaps the "most" non-linear equations are the Monge-Ampère equations, which are of importance in several geometrical problems. Great progress has been made in these areas in recent years.[14] With this heavy intrusion of analysis George Birkhoff's remark quoted above sounds even more disturbing. However, while analysis maps a whole mine, geometry looks out for the beautiful stones. Geometry is based on the principle that not all structures are equal and not all equations are equal.

Euler Characteristic as a Source of Global Invariants

To summarize, the Euler characteristic is the source and common cause of a large number of geometrical disciplines. I will illustrate this relationship by a diagram (see fig. 8).

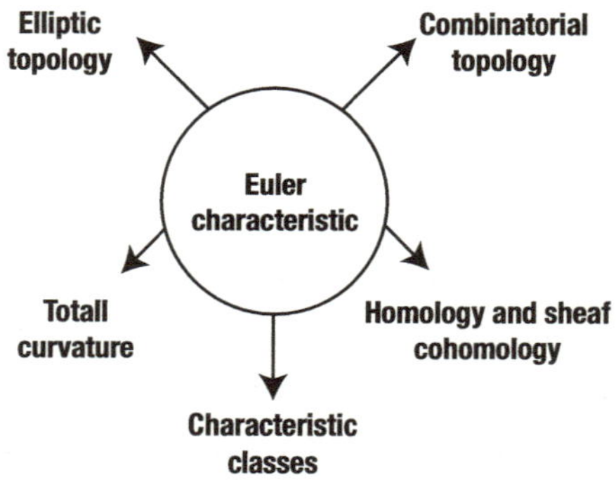

Fig. 8

Gauge Theory

At the beginning of this century differential geometry got the spotlight through Einstein's theory of relativity. Einstein's idea was to interpret physical phenomena as geometrical phenomena and to construct a space which would fit the physical world. It was a gigantic task and it is not clear whether he said the last word on a unified field theory of gravitational and electromagnetic fields. The introduction of vector bundles described above, and particularly the connections in them with their characteristic classes and their relations to curvature, widened the horizon of geometry. The case of a line bundle (i.e., when the fiber is a complex line) furnishes the mathematical basis of Weyl's gauge theory of an electromagnetic field. The Yang-Mills gauge fields, based on an understanding of the isotopic spin, is the first example of a nonabelian gauge theory. Its geometrical foundation is a complex plane bundle with a unitary connection. Attempts to unify all field theories, including strong and weak interactions, have recently focused on a gauge theory, i.e., a geometrical model based on bundles and connections. It is with great satisfaction to see geometry and physics united again.

Bundles, connections, cohomology, and characteristic classes are sophisticated concepts that crystallized after long years of search and experimentation in geometry. The physicist Chen-Ning Yang said: "That nonabelian gauge fields are conceptually identical to ideas in the beautiful theory of fiber bundles, developed by mathematicians without reference to the physical world, was a great marvel to me."[15] In 1975 he mentioned to me: "This is both thrilling and puzzling, since you mathematicians dreamed up these concepts out of nowhere." This puzzling is mutual. In fact, referring to the role of mathematics in physics, Eugene Wigner spoke about the unreasonable effectiveness of mathematics.[16] If one has to find a reason, it might be expressed in the vague term "unity of science." Fundamental concepts are always rare.

Concluding Remarks

Modern differential geometry is a young subject. Not counting the strong impetus it received from relativity and topology, its developments have been continuous. I am glad that we do not know what it is and, unlike many other mathematical disciplines, I hope it will not be axiomatized. With its contact with other domains in and outside of mathematics and with its spirit of relating the local and the global, it will remain a fertile area for years to come.

It may be interesting to characterize a period of mathematics by the number of variables in the functions or the dimension of the spaces it deals with. In this sense nineteenth century mathematics is one-dimensional and twentieth century mathematics is n-dimensional. It is because of the multi-variables that algebra acquires paramount importance. So far most of the global results on manifolds are concerned with even-dimensional ones. In particular, all complex algebraic varieties are of even real dimension. Odd-dimensional manifolds are still very mysterious. I venture to hope that they will receive more attention and substantial clarification in the twenty-first century. Recent works on hyperbolic three-manifolds by W. Thurston[17] and on closed minimal surfaces in a three-manifold by S. T. Yau, W. Meeks, and R. Schoen have thrown considerable light on three-manifolds and their geometry. Perhaps the problem of problems in geometry is still the so-called Poincaré conjecture which says that a closed simply connected three-dimensional manifold is homeomorphic to the three-sphere. Topological and algebraic methods have so far not led to a clarification of this problem. It is conceivable that tools in geometry and analysis will be found useful.

References

1. Veblen O, Whitehead J H C. Foundations of differential geometry. Cambridge; 1932. 17.
2. Cartan E. Le róle de la théorie des groupes de Lie dans l'évolution de la géométrie moderne. International Congress of Mathematicians. 1936; I: 96.
3. Birkhoff G D. Fifty years of American mathematics. Semi-Centennial addresses of Amer. American Mathematical Society. 1938: 307.
4. Weil A. S. S. Chern as friend and geometer. In: Chern. Selected Papers. Springer Verlag; 1978
5. Whitney H. Compositio Mathematica. 1937; 4: 276. (Max N. L. oversaw the production of the educational film *Regular Homotopy in the Plane* based on the Gauss-Bonnet-Whitney Theorem. For more details, refer to *American Mathematical Monthly*, 1978; 85: 212.—Trans.)
6. Pohl W F, Roberts G W. Journal of Mathematical Biology. 1978; 6: 383.
7. White J. H. American Journal of Mathematics. 1969; 91: 693; Fuller B. Proceedings of the National Academy of Sciences of the United States of America. 1971; 68: 815; Crick F. Proceedings of the National Academy of Sciences. 1976; 73: 2639.

8. Smale S. Transactions AMS. 1959; 90: 281; cf. also A. Phillips. Turning a surface inside out. Scientific American. May 1966; 214: 112–120. A film of the process, by N. L. Max, is distributed by the International Film Bureau, Chicago, Il.
9. Weyl H. Philosophy of mathematics and science. 1949. 90.
10. Einstein A. In library of living philosophers. I: 67. (The Chinese translation can be found in) The Collected Works of Einstein. 1977; 1: 30.
11. Hadamard J. Psychology of invention in the mathematical field. Princeton; 1945.115.
12. Godement R. Topologie algébrique et théorie des faisceaux. Paris: Hermann; 1958.
13. Chern S. Geometry of characteristic classes. Proc. 13th Biennial Sem. Canadian Mathematical Congress; 1972: 1.
14. Yau S. T. The róle of partial differential equations in differential geometry. International Congress of Mathematicians. Helsinki; 1978.
15. Yang C. N. Magnetic monopoles, gauge fields, and fiber bundles. Marshak Symposium; 1977. (The Chinese translation can be found in) Chinese Journal of Nature. 1979; 2 (10).
16. Wigner E. The unreasonable effectiveness of mathematics in the natural sciences. Communications on Pure and Applied Math. 1960; 13: 1.
17. Thurston W. Geometry and topology in dimension three. International Congress of Mathematicians. Helsinki; 1978.

Characteristic Classes and Characteristic Forms*

Lecture 1: Fiber Bundles and Characteristic Classes

▸ Introduction

It brings me great joy to witness the vibrant growth and diligent efforts of everyone here at the Institute of Mathematics of Academia Sinica. As Mr. Li Guowei mentioned earlier, this institute was initially established under my guidance. Allow me to share the story of its beginnings:

In 1946, I returned to Shanghai from the United States. Between 1943 and 1946, I had been conducting research on characteristic classes at the Institute for Advanced Study in Princeton, which will be the focus of today's lecture. Travel in those days was far from what it is today—it involved long journeys by ship. After more than twenty days at sea, I finally arrived in Shanghai, ready to teach at Tsinghua University. However, I was informed by the Academia Sinica that during the war, preparations had been made to establish an Institute of Mathematics. A preparatory office had been set up in Kunming, with Mr. Jiang Lifu serving as its director. So, once the institute was established, Mr. Jiang was going to be the director. But when the war ended, Mr. Jiang was sent abroad for further studies, and I was asked to assist in establishing the institute. I agreed to take on this responsibility.

---

* This article is divided into three lectures. The first two were delivered at Academia Sinica in Taiwan (October 19 and 26, 1990), and the third at National Taiwan University (November 2, 1990). The content was organized by Ming-Chang Kang and Hu Kai. It was originally published in *The Collected Works of Shiing-Shen Chern* by Linking Publishing Company (Taiwan) in 1993.

At the time, the preparatory office for the Institute of Mathematics was located in Shanghai. With funds returned from the Boxer Indemnity, the Institute of Natural Science had been established in the French Concession of Shanghai. The institute comprised several divisions, and we took over the mathematics section. This division was well-equipped, housing a complete collection of important mathematical journals of the time, such as *Math. Annalen*, *Crelle's Journal*, and *Liouville Journal*.

When I began my work, my most pressing task was training new talent. I assembled a group of about a dozen recent university graduates, most of whom had finished their studies within the last three years. What could they accomplish at that stage? I started delivering lectures in mathematics, as the majority of them—save for a few exceptions—had no idea how to embark on research. I thought that algebraic topology was an ideal starting point since it didn't require an extensive background. Thus, I gave six hours of lectures each week on algebraic topology, using *Topology* by Alexandrov and Hopf as the primary reference, which was considered a classic text at the time.

From this group of students, several went on to achieve remarkable success. For instance, Wu Wenjun later traveled to France and made significant contributions to the field of characteristic classes. Liao Shantao initially worked on fixed-point theorems for mappings on compact spaces before shifting his focus to dynamical systems. He is now at Peking University, and for his groundbreaking contributions to dynamical systems, he was awarded the Third World Academy of Sciences Mathematics Prize. Another distinguished figure is Mr. Yang Zhongdao, who is a fellow of the Academia Sinica.

Additionally, there was Chen Guocai, who made extraordinary contributions to differential geometry and Lie groups. Chen was an exceptionally original mathematician; his work in this field was more comprehensive and predates that of the renowned American mathematician Dennis Sullivan. Of course, Sullivan was also a brilliant thinker whose work spanned a broader range of topics.

▸ Stiefel-Whitney Classes and Pontryagin Classes

Today, we will discuss characteristic classes.

The importance of characteristic classes lies in their connection to a fundamental concept in modern mathematics: fiber bundles.

The simplest example of a fiber bundle can be explained as follows: You have probably studied analytic geometry and know how Descartes defined the coordinates of points in a plane. By selecting an x-axis and an arbitrary line l that intersects but does not coincide with the x-axis, all lines parallel to l cover the plane. If a line passing through a point P and parallel to l intersects the x-axis at a point with coordinate x, and P's coordinate along that line is y, then (x,y) represents the coordinates of P (fig. 1).

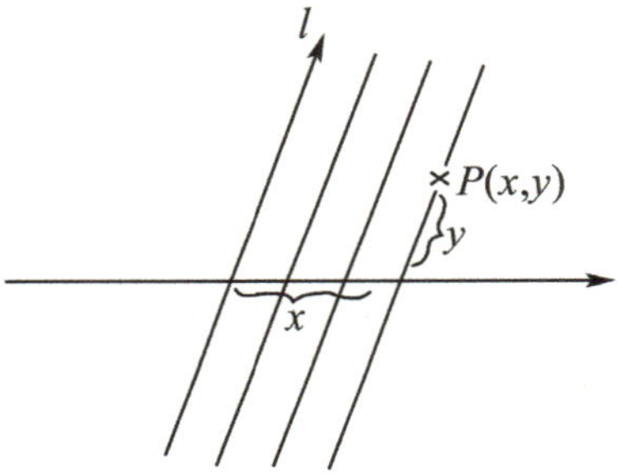

Fig. 1

Note that Descartes' definition of x and y is asymmetric. Later, mathematicians generalized this by making the axes orthogonal, introducing both an x-axis and a y-axis, which is a more specific case. In this setup, all lines parallel to l represent the fibers, and the entire plane can be viewed as the Cartesian product of the x-axis and l.

In general, a fiber bundle is a relaxation of this Cartesian product restriction. Instead of requiring the entire structure to be a product, we only require it to locally resemble a product. This relaxation is significant and vastly broadens the scope of fiber bundles to include many interesting cases. What is a local product?

Let E be a fiber bundle over a space M, with projection $\pi: E \longrightarrow M$. The term local product means that for every point x in M, there exists a neighborhood U of x such that $\pi^{-1}(U)$ is a product. Since $\pi^{-1}(U)$ is a product, points in this region have two kinds of coordinates: x, representing the coordinates of points in U; and y, representing coordinates on the fiber (fig. 2). This is entirely analogous to Descartes' setup.

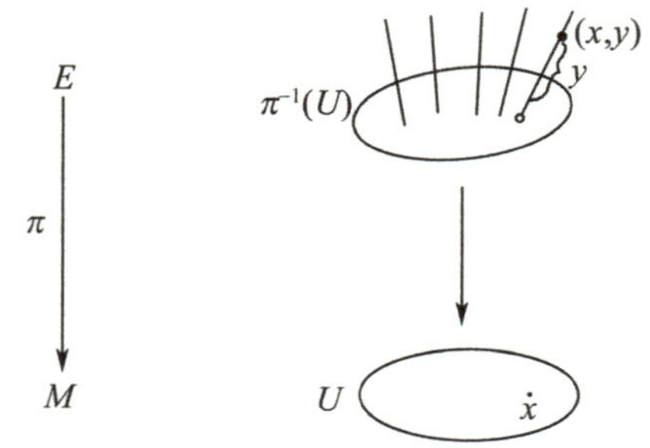

Fig. 2

However, the choice of the y-coordinates depends on the neighborhood U and is usually denoted y_U. If x has another neighborhood V such that $\pi^{-1}(V)$ is also a product, then points in $\pi^{-1}(x)$ will have two different coordinate representations: (x,y_U) and (x,y_V). The relationship between y_U and y_V is assumed to be a linear transformation: $y_U \cdot g_{UV} = y_V$, where g_{UV} is an element of the general linear group. This assumption ensures that the linear structure on the fibers is meaningful, giving us not just a general fiber bundle but a vector bundle.

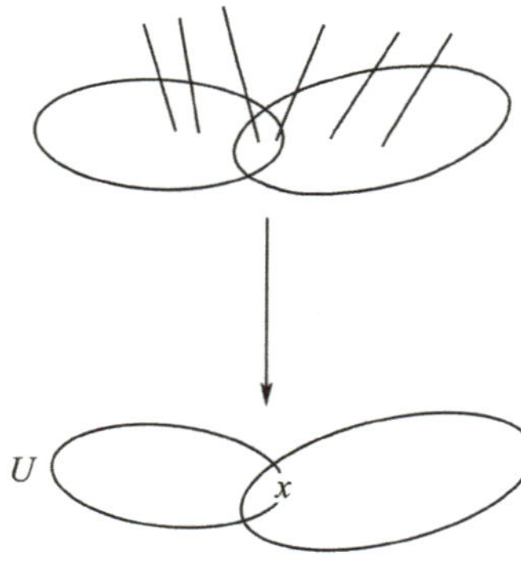

Fig. 3

More precisely, let M be a differentiable manifold, E a vector bundle over M, $x \in M$, and $\pi: E \longrightarrow M$ the projection. All spaces and maps are assumed to be smooth. Consider the following diagram:

$$\begin{array}{ccc} V_q & \xrightarrow{\ i\ } & E \\ {\scriptstyle \pi}\big\downarrow & & \big\downarrow \\ x & \longrightarrow & M \end{array} \tag{1}$$

Fig. 4

A vector bundle E is composed of vector spaces V_q, where $V_q = \pi^{-1}(x)$ and x is any point in M. Additionally, we require (1) for every point x in M, there exists a neighborhood U of x such that $\pi^{-1}(U)$ becomes a product manifold $U \times Y$, where Y is a q-dimensional vector space. Thus, the local coordinates of $\pi^{-1}(U)$ are (x, y_U), where $x \in U$, $y_U \in Y$; and (2) if two neighborhoods U and V intersect, then the coordinates y_U and y_V on the fibers satisfy the linear relationship:

$$y_U \cdot g_{UV}(x) = y_V \tag{2}$$

where $g_{UV}(x) \in GL(q)(= GL(q;\mathbf{R}) \text{ or } GL(q;\mathbf{C}))$ and the group $GL(q)$ acts from the right on Y.

Next, a natural question arises: Is a local product always a global product? This is not a trivial question, as it involves identifying invariants that characterize local products. This question was addressed by the **Poincaré-Hopf Theorem**:

Let M be a compact, oriented manifold, and let TM be its tangent bundle. Let $s: M \longrightarrow TM$ represent a generic vector field on M. Then

$$\chi(M) = \text{the total number of zeros of } s \tag{3}$$

where $\chi(M)$ is the Euler-Poincaré characteristic of M.

In the two-dimensional M case, Poincaré first proved this theorem, and Hopf later extended it to n-dimensional manifolds. If the tangent bundle TM is a global product, we can clearly construct a nowhere-vanishing vector field. Thus, if $\chi(M) \neq 0$, the Poincaré-Hopf Theorem implies that the tangent bundle cannot be a global product.

This is intriguing. Since the tangent bundle is not a global product, we can further investigate its unique properties.

The Poincaré-Hopf Theorem also has a deeper significance. We know that the Euler-Poincaré characteristic is a topological invariant, and the theorem tells us that the sum of zeros of a generic vector field is also a topological invariant. This insight can be extended further.

Since a single vector field defines the Euler-Poincaré characteristic, Hopf had an idea: What happens if we consider k vector fields instead of one? Suppose $s_1, s_2, \ldots, s_k$ are k vector fields in general position. The points in M where $s_1(x), s_2(x), \ldots, s_k(x)$ are linearly dependent form a trajectory of dimension $k - 1$.

Is this $(k - 1)$-dimensional trajectory a cycle (i.e., a closed chain in homology)? If so, is its homology class independent of the choice of vector fields? These questions are far from trivial because the $(k - 1)$-dimensional trajectory may not always form a $(k - 1)$-dimensional lower closed chain.

The work on these questions was conducted by E. Stiefel under the guidance of Hopf. Stiefel completed his doctoral dissertation in 1936. Around the same time, H. Whitney was also working in this area. While Stiefel made significant contributions, Whitney's contributions to characteristic classes far surpassed his.

During a period in the US, with Veblen's encouragement, algebraic topology flourished. However, I believe Whitney is arguably the greatest American algebraic topologist of the past decades. He was the first to recognize the importance of differential topology and laid its foundational theories.

Whitney made groundbreaking contributions to fiber bundles and characteristic classes, particularly by leveraging cohomology classes. He understood the significance of cohomology classes, not just as groups but also with their multiplicative structure, turning them into a ring—a critical development. Whitney also made fundamental contributions to stratified manifolds.

Stratified manifolds, in my opinion, hold a crucial place in the development of mathematics. In differential geometry, we typically assume manifolds are smooth, which allows us to use the tools of calculus. However, this assumption is unnatural—many practical examples, like tables and chairs, are not smooth. Stratified manifolds are more complex because they include singularities, and addressing singularities is mathematically challenging. Nevertheless, this progression is natural. Whitney wrote an excellent article on the development of topology in the American Mathematical Society's centennial historical volume, which is well worth reading.

Back to the original problem. We hoped that the $(k - 1)$-dimensional trajectory would form a closed chain. Whitney proposed that instead of treating it as a closed chain, we should seek an upper closed chain. The key is the cohomology class, and this idea is quite natural when you think about it because the condition for a vector to vanish, or for a set of vectors to be linearly dependent, is defined by a system of equations. These conditions are naturally expressed in terms of cohomology classes. Meanwhile, by using cohomology classes, the discussion is no longer restricted to tangent bundles; it can be extended to arbitrary vector bundles. How do we generalize the Euler-Poincaré characteristic? This generalization leads to the Stiefel-Whitney characteristic classes.

The idea is simple. Let $\pi: E \longrightarrow M$ be a fiber bundle over a manifold M. A mapping $s: M \longrightarrow E$ is called a section of E if it satisfies $\pi \cdot s = \mathrm{id}$. How do we find sections of a fiber bundle?

To construct a section, we divide the manifold M into small cells using a cellular decomposition. By refining the decomposition, we can ensure that the restriction of E to each cell becomes a topological product. Thus, to define a section of E, it suffices to specify its image on the fiber over each cell.

We proceed inductively, starting with the lowest-dimensional cells and extending to higher dimensions: Suppose the mapping's image is already defined at two vertices. Next, consider a one-dimensional cell connecting these two vertices. If the fiber is connected, the mapping can be easily extended along the edge. By induction, assume that the mapping is already defined on the boundary of a k-dimensional cell σ, i.e., on all $(k-1)$-dimensional cells. This means there exists a mapping $s: S^{k-1} \longrightarrow Y$, where Y is the fiber. Now, the problem is to extend s to σ. This is precisely a homotopy extension problem. If s represents an element of the homotopy group $\pi_{k-1}(Y)$ and is trivial (i.e., equal to zero), then s can be extended.

Thus, the obstruction to extending s only arises when the relevant homotopy group is nonzero. When the obstruction occurs, we not only stop but also obtain an invariant of the manifold itself. This dual advantage is powerful: When the homotopy group is zero, we can extend; when it is nonzero, we derive an invariant.

When considering k sections of a real vector bundle, the corresponding fiber is the Stiefel manifold $V_{q,k}$. A k-frame in $\mathbf{R}^q$ consists of k ordered, linearly independent vectors. The Stiefel manifold $V_{q,k}$ is the set of all k-frames in $\mathbf{R}^q$. Thus, the problem reduces to identifying the first nonzero homotopy group of $V_{q,k}$, or equivalently, its first nonzero cohomology group.

In the 1930s, computing the cohomology groups of specific spaces was far from straightforward. However, Stiefel accomplished this task, as did Whitney. For detailed results, I recommend Steenrod's classic book,[4] which remains a definitive reference in this field.

The culmination of these studies was the birth of characteristic classes: the Stiefel-Whitney classes of real vector bundles

$$\omega^k \in H^k(M, \mathbf{Z}_2), 1 \leq k \leq q \tag{4a}$$

and the Euler class of oriented bundles

$$\omega^q \in H^q(M, \mathbf{Z}) \tag{4b}$$

The reason for restricting the coefficients to $\mathbf{Z}_2$ in equation (4a) lies in the topological properties of the Stiefel manifold, since its cohomology group contains torsion elements.

Whitney had another idea, which involved the concept of the universal bundle. He generalized the tangent bundle to arbitrary vector bundles, and this generalization was groundbreaking.

Whitney observed that in a vector space of dimension $q + n$, all q-dimensional subspaces form a manifold of dimension qn. This is the Grassmannian manifold $Gr(q,n)$. On $Gr(q,n)$, there is already a natural vector bundle E_0. Why? Because a point on $Gr(q,n)$ is itself a vector space, and we simply treat it as the fiber. The vector bundle associated with this is

$$\pi_0 : E_0 \longrightarrow Gr(q,n) \tag{5}$$

This is called the tautological bundle, also known as universal bundle because any vector bundle can be induced from it by a map, $f: M \longrightarrow Gr(q,n)$, provided n is large enough.

Similarly, Pontryagin discovered, when n is large enough, that any map f homotopic to the above generates the same vector bundle. This leads to the natural definition of characteristic classes as elements in the following set:

$$f^* H^*(Gr(q,n)) \subset H^*(M)$$

Thus, the importance of the Grassmannian manifold is clear. Its topological structure is closely related to intersection theory in algebraic geometry, a fundamental problem. The person who made the most important contributions to this topological problem was C. Eresmann. His doctoral dissertation, completed in 1934, began the study of this issue. After determining the structure of the homology groups of the Grassmannian manifold, one can compute the characteristic classes of any vector bundle over any manifold. When the coefficients are $\mathbf{Z}_2$, we obtain the Stiefel-Whitney classes. However, we are particularly interested in the cases where the coefficients are integer coefficients or rational coefficients.

Pontryagin, as well as Eresmann, realized that, aside from the Euler class in equation (4b), the integer-valued characteristic classes of real vector bundles appear only in dimensions $4k$. These are the Pontryagin classes, denoted by

$$p_k(E) \in H^{4k}(M, \mathbf{Z})$$

These developments regarding characteristic classes were primarily completed around 1940.

I arrived at the Institute for Advanced Study in Princeton in 1943. My starting point was the **Gauss-Bonnet Theorem** from differential geometry, which connects curvature to the Euler-Poincaré characteristic:

Let M be a closed oriented surface, K its Gaussian curvature, and $\mathrm{d}A$ the area differential. Then, we have

$$\frac{1}{2\pi} \int K \mathrm{d}A = \chi(M) \tag{6}$$

In fact, neither Gauss nor Bonnet ever wrote down the exact form of the equation (6). Gauss was considering a geodesic triangle of constant curvature, as shown in fig. 5.

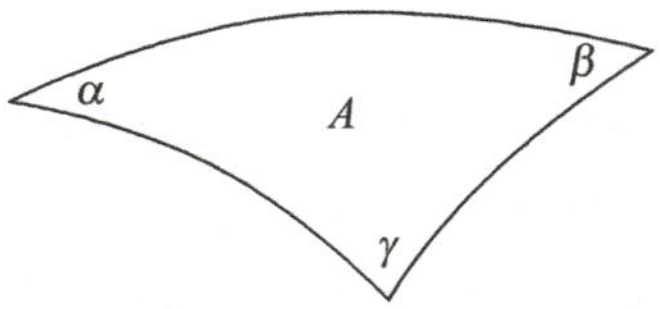

Fig. 5

Gauss proved $\alpha + \beta + \gamma - \pi = KA$.

Bonnet relaxed the restriction to geodesic curves, extending the result to arbitrary curves, but included their geodesic curvatures as corrections.

It took a long time to progress from the formulas of Gauss and Bonnet to the modern formulation (equation 6). This delay was partly due to the lack of clarity in the 19th century regarding concepts like the Euler-Poincaré characteristic and triangulations. At the time, different researchers interpreted these ideas in disconnected ways. However, among these concepts, "genus" stood out as an important idea.

The genus of a closed oriented surface is a topological invariant that counts the number of "holes" in the surface. If g is the genus, then $\chi = 2 - 2g$.

On the other hand, any oriented surface has a complex structure, allowing the study of holomorphic differentials on it. A key theorem in algebraic curve theory states that the space of holomorphic differentials on a closed oriented surface forms a complex vector space of dimension g. This is a profound result, as it connects topological invariants with complex structure invariants. To extend these results to higher dimensions, characteristic classes serve as a powerful tool.

In 1943, I met André Weil at Princeton.

During World War II, after France entered the war and began conscription, Weil—who was a pacifist—fled abroad. He was arrested by the French government and imprisoned. The preface of his book *L'intégration dans les groupes topologiques et ses applications* was written while he was in prison. Shortly after, France surrendered, and Weil was released. When Weil came to the United States, the war caused institutions to tighten budgets, making it hard for him to find a job. He ended up teaching calculus at Lehigh University, which he found quite painful.

At the time, Weil had just coauthored a significant paper with Carl Allendoerfer on a Gauss-Bonnet formula for polyhedra. This was a remarkable and groundbreaking paper because polyhedra have singularities, making the discussion of integration on them highly complex. Weil spoke to me about why polyhedra presented such difficulties.

Based on my understanding of the two-dimensional case, I began exploring how to generalize it to higher dimensions. In two dimensions, the Gauss-Bonnet formula connects the Euler-Poincaré characteristic with curvature. This led me to consider how, in higher dimensions, one could relate curvature to Stiefel-Whitney characteristic classes.

One Sunday, I went to my office. Although Princeton's mathematics building was closed on Sundays, every mathematician had a master key to the building. I wandered into the library, casually browsing, and suddenly had an idea: Why not try using complex vector bundles?

Some vector bundles lack an inherent complex structure, but their fibers still possess one. Historically, Stiefel-Whitney and Pontryagin classes were limited to real coefficients because the cohomology groups of Stiefel and Grassmann manifolds contain torsion elements. This restriction meant the classes could only be considered in $\mathbf{Z}_2$-cohomology. However, the situation simplifies significantly when working with complex Stiefel and Grassmann manifolds.

So I extended Stiefel and Whitney's methods to complex coefficients and complex vector bundles. This work resulted in the Chern class, defined as

$$c_k(E) \in H^{2k}(M, \mathbf{Z})$$

The Chern classes are the most important characteristic class. The coefficients of the Stiefel-Whitney classes in $\mathbf{Z}_2$ are not very useful, and the Pontryagin classes either. The Chern classes, however, manifest locally as curvature and are globally integral cohomology classes, making them the most widely applicable characteristic classes.

Lecture 2: Chern Classes, Chern Forms, and Transgression

▸ Characteristic Classes and Curvature

Among all the characteristic classes, I personally believe the Chern classes are the most important.

Now, consider a complex vector bundle E over a manifold M. If E is a q-dimensional complex vector bundle over M, we define a series of invariants for E, the Chern classes $c_k(E)$ as $c_k(E) \in H^{2k}(M, \mathbf{Z})$, $1 \leq k \leq q$.

When M is a q-dimensional compact complex manifold and E its holomorphic tangent bundle, the top Chern class $c_q(E)$ lies in the highest-dimensional cohomology group. Moreover,

$$\int_M c_q(E)$$

is an integer, which actually equals the Euler-Poincaré characteristic of M. Thus, Chern classes naturally generalize the Euler-Poincaré characteristic. We will define the Chern classes using three different approaches: the obstruction theory approach, the universal bundle approach, the curvature approach.

The obstruction theory approach. Similar to the definition of Stiefel-Whitney classes for real vector bundles in the previous lecture, we consider a cell decomposition of the complex manifold M, but now focus on the complex Stiefel manifold $V_{q,k}(\mathbf{C})$.

For a q-dimensional complex vector bundle E, where $1 \leq k \leq q$, if Σ_k represents the set of all k-dimensional cells in the cell decomposition of M, we obtain a map:

$$\gamma_k : \Sigma_{2k} \longrightarrow \pi_{2k-1}(V_{q,q-k+1}(\mathbf{C}))$$

It can be proven that γ_k forms an upper closed chain, and its cohomology class does not depend on the choice of these k-sections. We denote this class by $c_k(E)$.

Since the homotopy groups of $V_{q,k}(\mathbf{C})$ have a long sequence of zeros,

$$\pi_i(V_{q,q-k}(\mathbf{C})) = \begin{cases} 0, \text{if } 0 \leq i \leq 2k \\ \mathbf{Z}, \text{if } i = 2k+1 \end{cases}$$

we have $c_k(E) \in H^{2k}(M, \mathbf{Z})$. This represents the obstruction to finding $q - k + 1$ linearly independent sections of E.

Additionally, for the real Stiefel manifold $V_{q,k}(\mathbf{R})$, its homotopy groups are

$$\pi_i(V_{q,q-k}(\mathbf{R})) = \begin{cases} 0, \text{if } 0 \leq i \leq k-1 \\ \mathbf{Z}_2, \text{if } i = k \text{ is odd and } 1 \leq k \leq q-2 \\ \mathbf{Z}, \text{if } i = k \text{ is even or } k = q-1 \text{ or } q \end{cases}$$

Regardless, there always exists a non-zero homomorphism:

$$\pi_k(V_{q,q-k}(\mathbf{R})) \longrightarrow \mathbf{Z}_2$$

which is why Stiefel-Whitney classes for real vector bundles are cohomology classes with coefficients in $\mathbf{Z}_2$.

The universal bundle approach. This is similar to the universal bundle approach discussed in the previous lecture, but now we consider the complex Grassmannian $Gr(q, n; \mathbf{C})$.

In 1934, Ehresmann used Schubert cells from algebraic geometry to perform a cell decomposition of $Gr(q, n; \mathbf{C})$. Let σ_1, σ_2, …, σ_q be a set of integers and L_1, L_2, …, Lq be a set of subspaces of $\mathbf{C}^{q+n}$ such that

$$\begin{cases} 0 \leq \sigma_1 \leq \sigma_2 \leq \cdots \leq \sigma_q \leq n \\ \dim L_i = \sigma_i + i \end{cases}$$

Since each point $[W]$ in $Gr(q,n;\mathbf{C})$ represents a q-dimensional subspace W of C^{q+n}, we define the Schubert symbol $(\sigma_1, \sigma_2, \ldots, \sigma_q)$ as

$$(\sigma_1,\sigma_2,\ldots,\sigma_q)=\{[W]\in Gr(q,n;\mathbf{C})\colon \dim(W\cap L_i)\geq i\}$$

As σ_1, σ_2, ..., σ_q vary, these Schubert symbols form a cell decomposition of $Gr(g,n;\mathbf{C})$, and they are still lower closed chains, representing a homology class.

Note that for real coefficients, Schubert symbols may not form lower closed chains in $Gr(q,n;\mathbf{C})$. We use the same notation $(\sigma_1, \sigma_2, \ldots, \sigma_q)$ to denote the dual cohomology class of these Schubert symbols. These Schubert symbols form a basis for the homology group $H_+(Gr(q,n;\mathbf{C}),\mathbf{Z})$. Thus, the pairing of cohomology and homology classes becomes

$$\langle(\sigma_1,\ldots,\sigma_q),(\tau_1,\ldots,\tau_q)\rangle=\begin{cases}0,(\sigma_1,\ldots,\sigma_q)\neq(\tau_1,\ldots,\tau_q)\\1,(\sigma_1,\ldots,\sigma_q)=(\tau_1,\ldots,\tau_q)\end{cases}$$

Among the cohomology classes represented by these Schubert symbols, there is one particularly basic cohomology class:

$$\sigma^{(k)}=(0,\ldots,0,1,\ldots,1)\in H^{2k}(Gr(q,n;\mathbf{C}),\mathbf{Z})$$

where there are k ones and $q-k$ zeros. When n is sufficiently large, $\sigma^{(k)}$ is the k-th Chern class of the universal bundle over $Gr(q,n;\mathbf{C})$.

For any q-dimensional complex vector bundle E over a manifold M, we first choose a sufficiently large integer n and an appropriate map f, $f\colon M\longrightarrow Gr(q,n;\mathbf{C})$, such that $E=f^*(E_0)$, where E_0 is the universal bundle over $Gr(q,n;\mathbf{C})$. We then define the k-th Chern class of E as $c_k(E)=f^*(\sigma^{(k)})\in H^{2k}(M,\mathbf{Z})$.

The curvature approach. In my opinion, the most useful approach is the one based on curvature.

Let E be a q-dimensional complex vector bundle over the manifold M, and let $\Gamma(E)$ be the space of sections of E, which forms a complex vector space.

We want to differentiate elements of $\Gamma(E)$, and for this, we need the concept of a "connection." A connection is a map:

$$D\colon\Gamma(E)\longrightarrow\Gamma(E\otimes T^*M) \tag{7}$$

where T^*M is the complex cotangent bundle of M, and it must satisfy the following conditions:

$$\text{(D1)}: D(s_1+s_2)=D(s_1)+D(s_2) \tag{8}$$

$$\text{(D2)}: D(f(x)s)=\mathrm{d}f\otimes s+f\cdot Ds \tag{9}$$

where $s,s_1,s_2 \in \Gamma(E)$ and $f(x)$ is a smooth function on M. This kind of connection exists not only for complex vector bundles but also for real-analytic vector bundles, where the connection can be real-analytic.

Now, we proceed to derive the local representation of the connection. First, we define a frame: a set of q ordered sections $s_1,s_2,\ldots,s_q$ such that $s_1\wedge\ldots\wedge s_q$ is nowhere zero in a neighborhood U, i.e., $s_1,\ldots,s_q$ form a frame in U.

Let $s_1,\ldots,s_q$ be a frame of the vector bundle E over U. We write D_{si} as

$$Ds_i=\sum\omega_i^j\otimes s_j, 1\le i,j\le q \tag{10}$$

where ω_i^j are first-order differential forms, and let

$$\omega=(\omega_i^j) \tag{11}$$

which is called the connection matrix. Using a simplified matrix notation, equation (10) can be written as

$$Ds=\omega s \tag{12}$$

Since any section can be written as a linear combination of the frame sections, once we have ω, the connection D is completely determined.

If $s_1',\ldots,s_q'$ is another frame over U, then how is the connection matrix ω determined by $s_1,\ldots,s_q$ related to the connection matrix ω' determined by $s_1',\ldots,s_q'$? Let

$$s'(x)=g(x)s(x), x\in U \tag{13}$$

where $g(x)$ is a non-singular $q\times q$ matrix, and the elements of $g(x)$ are smooth functions on U. Differentiating equation (13) using (D1) and (D2), we obtain

$$dg+g\omega=\omega' g \tag{14}$$

Now, taking the exterior derivative of equation (14), we get

$$g\Omega=\Omega' g \tag{15}$$

where Ω and Ω' are defined as

$$\Omega=d\omega-\omega\wedge\omega, \Omega'=d\omega'-\omega'\wedge\omega \tag{16}$$

According to the definition of Ω, it is a $q \times q$ matrix, and its elements are two-forms. Ω is called the curvature matrix.

Equation (15) leads us to consider the following determinant:

$$\det\left(I+\frac{i}{2\pi}\Omega\right)=1+c_1(\Omega)+\ldots+c_q(\Omega) \tag{17}$$

where $c_k(\Omega)$ is a $2k$-form defined on U (for $1 \le k \le q$). The value of $c_k(\Omega)$ is independent of the choice of the frame $s_1,\ldots,s_q$.

Now let $\{U_a\}$ be an open cover of the manifold M. Define $c_k(\Omega)$ on each open set U_a. Equations (15) and (17) imply that U_a is consistent on the overlaps of any two open sets. Thus, $c_k(\Omega)$ is a globally defined $2k$-form on M. These forms $c_k(\Omega)$ are actually closed. By de Rham Theorem, $c_k(\Omega)$ corresponds to a cohomology class: $c_k(E) \in H^{2k}(M,\mathbf{R})$.

The forms $c_k(\Omega)$ are called the Chern forms. It can be proven that the Chern classes $c_k(E)$ defined using Chern forms are consistent with those defined using obstruction theory or the universal bundle. This is achieved by using the universal bundle, as all the above calculations commute with the map $f{:}M \longrightarrow Gr(q,n;\mathbf{C})$, where $Gr(q,n;\mathbf{C})$ is the Grassmannian manifold. So, we take this comparison and observe it on the universal bundle over $Gr(q,n;\mathbf{C})$. But on the universal bundle, all quantities can be explicitly and clearly described.

By complexifying a real vector bundle, one can express Pontryagin classes in terms of Chern classes. This curvature-based definition of $c_k(E)$ has led to significant developments later; refer to [1] and [2] for more details.

In F. Hirzebruch's *Topological Methods in Algebraic Geometry*, another method for defining Chern classes is given. It determines characteristic classes through a set of axioms. While this method is concise and encompasses various characteristic classes, sufficient for the issues that Herr Schieblich wants to apply, it lacks an explanation of the geometric meaning of characteristic classes.

Let's return to characteristic classes of complex vector bundles: Let E_1 and E_2 be complex vector bundles of dimensions q_1 and q_2, respectively, over a manifold M. Consider the Whitney sum $E_1 \oplus E_2$. How can the characteristic classes of E_1 and E_2 express the characteristic classes of $E_1 \oplus E_2$? Whitney provided a convenient formula for this, known as the Whitney's Duality Theorem.

Whitney's original proof was exceedingly lengthy, spanning over 100 pages. As mentioned in my previous lecture, in a historical essay in the AMS centennial collection, Whitney stated that later R. Thom found a new method. This is incorrect. The new method was discovered by Wu Wenjun, and his work was published in *Ann. Math.* When Wu first met me, he did not get topology, but within three years, he had provided a simple proof of Whitney's Duality Theorem. Whitney's original proof was never published.

Let us sum the Chern classes of a complex vector bundle E,

$$c(E) = 1 + c_1(E) + \ldots + c_q(E) \in H^*(M, \mathbf{Z})$$

where $H^*(M, \mathbf{Z})$ is the cohomology ring of M with cup product. Whitney's Duality Theorem states $c(E_1 \oplus E_2) = c(E_1) \cdot c(E_2)$.

We write $c(E)$ as

$$c(E) = \prod_{i=1}^{q} \{1 + \gamma_i(E)\}$$

where are formal variables, and the elementary symmetric polynomials of are the Chern classes $c_k(E)$ (for $1 \le k \le q$). Now define the Chern character:

$$\mathrm{ch}(E) = \sum_{i=1}^{q} e^{\gamma_i(E)}$$

The advantage of the Chern character is that it transforms multiplication into addition. Thus, we have

$$\mathrm{ch}(E_1 \oplus E_2) = \mathrm{ch}(E_1) + \mathrm{ch}(E_2), \mathrm{ch}(E_1 \otimes E_2) = \mathrm{ch}(E_1) \cdot \mathrm{ch}(E_2)$$

where $E_1 \otimes E_2$ is the tensor product of two vector bundles.

Now, it is well-known that topology, analysis, and operator algebras have undergone one significant development, the K-theory. By collecting all complex vector bundles (of any dimension) on a manifold M, applying appropriate equivalence relations (using the Grothendieck construction), and defining addition as the Whitney sum of bundles and multiplication as the tensor product, we obtain a commutative ring $K(M)$.

The two properties of the Chern character above can be restated as the Chern map, which is a ring homomorphism from the K-theory ring $K(M)$ to the cohomology ring $H^*(M, \mathbf{Z})$: $\mathrm{ch}: K(M) \longrightarrow H^*(M, \mathbf{Z})$.

Atiyah and Hirzebruch proved that if M is a finite CW complex, then

$$\mathrm{ch}: K(M) \otimes \mathbf{Q} \longrightarrow H^{2\bullet}(M, \mathbf{Q})$$

is an isomorphism.

The Chern character also plays an essential role in index theory. Let $D: E \longrightarrow F$ be an elliptic operator between vector bundles E and F. The **Atiyah-Singer Index Theorem** states as follows:

Let ind(D) = dim(Ker D) − dim(Coker D), where Td(M) is the Todd class of the oriented compact manifold M. Then

$$\mathrm{ind}(D) = \int_M \mathrm{ch}(D) \cdot \mathrm{Td}(M)$$

Please note that we have only defined the Chern character of complex vector bundles, not that of elliptic operators. Nevertheless, this equation's importance is evident because the left-hand side is an analytic index, while the right-hand side is a geometric index. This is a global result, exceptionally deep and important.

▸ Transgression

We mentioned earlier that the Chern classes $c_k(E)$ determined by the Chern forms $c_k(\Omega)$ are invariants of the vector bundle E, and thus they are independent of the choice of connection D. In fact, this can be directly proven: if D and D' are connections on the complex vector bundle E, and Ω and Ω' are the curvature forms associated with D and D', respectively, then there exists a $2k-1$-form Q defined globally on the manifold M such that $c_k(\Omega) - c_k(\Omega') = \mathrm{d}Q$.

This process is referred to as transgression.

We can also illustrate the phenomenon of transgression using the proof of the Gauss-Bonnet Theorem.

Let M be a two-dimensional oriented Riemannian manifold, and $x \in M$. For any unit tangent vector e_1 at x, there exists a unique unit tangent vector e_2 such that e_1 and e_2 are orthogonal and form an oriented basis consistent with the orientation of M. Let ω_1 and ω_2 denote the dual frames of e_1 and e_2, which are defined on the unit tangent bundle E over M. The projection

$$\pi: E \longrightarrow M \tag{18}$$

maps E to M. In fact, $\omega_1 \wedge \omega_2$ is a differential form on M and serves as the area element of M.

Define a connection form ω_{12} as the unique one-form satisfying

$$d\omega_1 = \omega_{12} \wedge \omega_2,\ d\omega_2 = \omega_1 \wedge \omega_{12} \tag{19}$$

Equation (19) indicates that the connection has no torsion.

Taking the exterior derivative of equation (19) yields

$$d\omega_{12} = -K\omega_1 \wedge \omega_2 \tag{20}$$

where K is the Gaussian curvature.

Equation (20) is extremely important because the Gauss-Bonnet Theorem states that

$$\frac{1}{2\pi} K\omega_1 \wedge \omega_2$$

is precisely the Euler class. Moreover, the equation shows that when this is pulled back to E, it becomes integrable, and its integral equals the connection form. This provides the most natural proof of the Gauss-Bonnet Theorem, which can be extended to higher dimensions and has numerous other applications.

The phenomenon of transgression also occurs in principal bundles or frame bundles. We illustrate this using the Chern-Simons form.

In equation (13), if g is any nonsingular $q \times q$ matrix, we obtain various possible frames over M. Thus, the local coordinates of the frame bundle P can be written as (x,g).

On the frame bundle P, let φ be a connection form. According to equation (14), the local expression for φ is

$$\varphi = dg \cdot g^{-1} + g\omega g^{-1} \tag{21}$$

Hence, the following curvature form is also defined on P:

$$\Phi = d\varphi - \varphi \wedge \varphi = g\Omega g^{-1} \tag{22}$$

From equation (17), we find

$$c_1(\Phi) = \frac{i}{2\pi}\mathrm{Tr}\Phi$$

Since $\mathrm{Tr}(\varphi < \varphi) = 0$, according to equation (22), we have

$$c_1(\Phi) = \frac{i}{2\pi} d\mathrm{Tr}\varphi \tag{23}$$

In other words, on the frame bundle P, the first Chern form $c_1(\Phi)$ becomes an exact differential form.

We now define the Chern-Simons form.

Similarly, from equation (17), we have

$$c_2(\Phi) = \left(\frac{i}{2\pi}\right)^2 \{c_1(\Phi)^2 - \mathrm{Tr}(\Phi \wedge \Phi)\} \tag{24}$$

Taking the exterior derivative of the curvature equation yields the Bianchi identity, we have

$$\mathrm{d}\Phi = \varphi \wedge \Phi - \Phi \wedge \varphi \tag{25}$$

From equation (22) and (25), we derive

$$\mathrm{d}\{\mathrm{Tr}(\varphi \wedge \varphi \wedge \varphi)\} = 3\mathrm{Tr}(\varphi \wedge \varphi \wedge \Phi)$$
$$\mathrm{d}\mathrm{Tr}(\varphi \wedge \Phi) = -\mathrm{Tr}(\varphi \wedge \varphi \wedge \Phi) + \mathrm{Tr}(\Phi \wedge \Phi)$$

Let

$$CS(\varphi) = \frac{1}{3}\mathrm{Tr}(\varphi \wedge \varphi \wedge \varphi) + \mathrm{Tr}(\varphi \wedge \Phi) \tag{26}$$

It is straightforward to verify

$$\mathrm{d}CS(\varphi) = \mathrm{Tr}(\Phi \wedge \Phi) \tag{27}$$

The differential form defined in equation (26) is called the Chern-Simons form. It is the transgressed differential form of the second Chern form $c_2(\Phi)$ on the frame bundle. The Chern-Simons form is itself a three-form, and its definition does not depend on the metric of M. When M is a three-dimensional manifold, the restriction on the dimension implies that it must be a closed differential form. Recently, the Chern-Simons form has played an important role in theoretical physics, please see [6].

Professor Chen-Ning Yang recently told me that we are all very fortunate. The Yang-Mills gauge fields, the Yang-Baxter equation, and now the Chern-Simons form are all enjoying significant attention. Some scientists may make important contributions, and while their work is recognized, their work ends, and leads to further developments. However, for his situation, mine, and the Chern-Simons form, many researchers are currently working on these topics.

When I was at Princeton, I met many topologists. Except for me, they were all influenced by Steenrod, and they insisted on using upper closed chains instead of differential forms. However, with upper closed chains, handling multiplication becomes difficult, whereas differential forms make it much easier. Yet differential forms were not the trend.

So I think this is also a lesson—do not follow the crowd. Back at Hamburg University, many people were studying number theory. Artin, Hecke, and Petersen were all excellent number theorists, and I was friends with all of them. However, I chose not to study it, even though I found it very interesting. Research must be original—do not follow others.

Lecture 3: Complex Line Bundles and Holomorphic Bundle Theory

▸ Holomorphic Line Bundles and the Nevanlinna Theory

Vector bundles are a rich and complex mathematical concept. When the fibers are one-dimensional, they are called line bundles. Complex line bundles have enormous importance in many areas of mathematics and theoretical physics. In this case, the fibers are $\mathbf{C}$, and the structure group is $GL(1,\mathbf{C}) = C^* = \mathbf{C}\backslash\{0\}$, which is an Abelian group. Therefore, one can define addition between line bundles, resulting in the group of complex line bundles.

Let's first examine the relationship between the cohomology group of sheaves and the Chern classes of complex line bundles.

By the definition of the sheaf cohomology group, the group of complex line bundles is isomorphic to $H^1(M,C_M^*)$, where C_M is the sheaf of complex-valued functions on M and C_M^* is the sheaf of complex-valued functions that do not take the value zero.

Consider the following exact sequence, $0 \longrightarrow \mathbf{Z} \longrightarrow C_M \xrightarrow{e} C_M^* \longrightarrow 0$, where the map e is defined by $e(f) = \exp(2\pi i f)$.

Thus, we obtain a long exact sequence,

$$\ldots \longrightarrow H^1(M,C_M) \longrightarrow H^1(M,C_M^*) \xrightarrow{c} H^2(M,\mathbf{Z}) \longrightarrow \ldots$$

where the map c maps a complex line bundle to the function given by its first Chern class.

Since C_M is a fine sheaf, its non-zero-dimensional cohomology groups are all zero. From the above exact sequence, we get $H^1(M, C_M^*) \cong H^2(M, \mathbf{Z})$.

This relationship gives a geometric interpretation of the second cohomology group with integer coefficients: It is isomorphic to the group of complex line bundles.

If M is a complex manifold and all maps are holomorphic, we then obtain the concept of holomorphic line bundles. The group of holomorphic line bundles is isomorphic to $H^1(M, O_M^*)$, where O_M^* is the sheaf of non-zero holomorphic functions on M. This structure is more intricate, and its properties are much richer. It generalizes the concept of divisors and is fundamental to the structure of complex manifolds.

Now, we adopt the perspective of the universal bundles.

Since our vector bundle is one-dimensional, the base space of the universal bundle is $Gr(1,n;\mathbf{C})$. Note that $Gr(1,n;\mathbf{C}) = P^n(\mathbf{C})$, which is the complex projective space and the universal bundle pulls out the zero section, giving $\mathbf{C}^{n+1}\backslash\{0\}$, still denoted by π_0:

$$\pi_0 : \mathbf{C}^{n+1}\backslash\{0\} \longrightarrow P^n(\mathbf{C}) \tag{28}$$

Let $Z = (x_0, x_1, \ldots, x_n) \in \mathbf{C}^{n+1}\backslash\{0\}$, and the value of $\pi_0(Z)$ is

$$\pi_0(Z) = [x_0 : x_1 : \ldots : x_n] \in P^n(\mathbf{C})$$

where $[x_0 : x_1 : \ldots : x_n]$ represents the homogeneous coordinates of $P^n(\mathbf{C})$.

If we restrict the map π_0 to the unit sphere S^{2n+1},

$$S^{2n+1} = \{Z \in \mathbf{C}^{n+1} \mid x_0\overline{x_0} + \ldots + x_n\overline{x_n} = 1\}$$

we obtain the Hopf fibration of S^{2n+1}.

When $n = 1$, $\pi_0 : S^3 \longrightarrow P^1(\mathbf{C}) = S^2$, which is the famous Hopf fibration. It is the first null-homotopic map from a higher-dimensional space to a lower-dimensional space.

Returning to the universal bundle of $P^n(\mathbf{C})$, we first consider the cohomology groups of $P^n(\mathbf{C})$.

Since

$$H_i(P^n(\mathbf{C}),\mathbf{Z}) = \begin{cases} 0, \text{ when } i \text{ is odd} \\ \mathbf{Z}, \text{ when } i \text{ is even and } i \leq 2n \end{cases}$$

In fact, the projective subspace $P^k(\mathbf{C})$ of $P^n(\mathbf{C})$, where $1 \leq k \leq n$, is a closed cycle and generates $H_{2k}(P^n(\mathbf{C}),\mathbf{Z})$.

Let $\xi \in H^2(P^n(\mathbf{C}),\mathbf{Z})$ be the dual cohomology class of the $2n-2$-dimensional closed cycle $P^{n-1}(\mathbf{C})$. This ξ is the first Chern class of the universal bundle over $P^n(\mathbf{C})$. A nice feature of ξ is that the dual cohomology class of the closed cycle of dimension $2k$, $P^k(\mathbf{C})$, is given by $\xi^{n-k} \in H^{2n-2k}(P^n(\mathbf{C}),\mathbf{Z})$.

If E is a complex line bundle on a manifold M, and the map $f{:}M \longrightarrow P^n(\mathbf{C})$ satisfies $E = f^*(E_0)$, where E_0 is the universal bundle over $P^n(\mathbf{C})$, then the Chern class of E is $c_1(E) = f^*(\xi) \in H^2(M,\mathbf{Z})$.

Now, let's return to the geometric properties of the universal bundle in equation (28). We need to define its connection and curvature. Let $Z, W \in \mathbf{C}^{n+1}$ be given by

$$Z = (x_0, x_1, \ldots, x_n), W = (y_0, y_1, \ldots, y_n) \in \mathbf{C}^{n+1} \tag{29}$$

and define their inner product by

$$\langle Z, W\rangle = x_0\overline{y}_0 + x_1\overline{y}_1 + \ldots + x_n\overline{y}_n \tag{30}$$

A set of vectors $Z_0, Z_1, \ldots, Z_n \in \mathbf{C}^{n+1}$ satisfies the condition

$$\langle Z_A, Z_B\rangle = \delta_{A\overline{B}}, 0 \leq A, B \leq n \tag{31}$$

then $Z_0, Z_1, \ldots, Z_n$ form an orthonormal frame.

All such orthonormal frames correspond one-to-one to elements of the unitary group $U(n+1)$.

We define the Maurer-Cartan form $\omega_{A\overline{B}}$ as $\omega_{A\overline{B}} = \langle \mathrm{d}Z_A, Z_B\rangle$.

Thus, the differential $\mathrm{d}Z_A$ can be written as

$$\mathrm{d}Z_A = \sum_{B=0}^{n} \omega_{A\overline{B}} Z_B \tag{32}$$

Differentiating equation (31) gives

$$\omega_{A\overline{B}} + \omega_{\overline{B}A} = 0, \text{ where } \omega_{\overline{B}A} = \overline{\omega_{B\overline{A}}} \tag{33}$$

Now, applying exterior differentiation to equation (32), we get

$$\mathrm{d}\omega_{A\bar{B}} = \sum_{C=0}^{n} \omega_{A\bar{C}} \wedge \omega_{C\bar{B}} \tag{34}$$

which is the Maurer-Cartan equation.

Now how to define the connection on the universal bundle over $P^n(\mathbf{C})$? We just need to specify it locally.

For any $Z_0 \in \mathbf{C}^{n+1}$ with $\langle Z_0, Z_0 \rangle = 1$, we can view Z_0 as a point in $P^n(\mathbf{C})$, and its fiber is the line represented by Z_0. Once we know how to differentiate Z_0, we can deduce the differentiation of all other points on its fiber.

For $Z_0, Z_1, \ldots, Z_n \in \mathbf{C}^{n+1}$ that form an orthonormal frame of $Z_0, Z_1, \ldots, Z_n$, we define the connection, following the method of Levi-Civita used before:

$$DZ_0 = \omega_{0\bar{0}} Z_0 \tag{35}$$

From equation (34), the curvature of connection D is

$$\Omega = \mathrm{d}\omega_{0\bar{0}} = \sum_{k=0}^{n} \omega_{0\bar{k}} \wedge \omega_{k\bar{0}} - \sum_{k=0}^{n} \omega_{0\bar{k}} \wedge \omega_{\bar{0}k} \tag{36}$$

Please note that equation (36) represents the transgression on the universal bundle because its right-hand side is a second-order differential form on $P^n(\mathbf{C})$.

In another direction,

$$\mathrm{d}s^2 = \sum_{k=0}^{n} \omega_{0\bar{k}} \omega_{0\bar{k}} \tag{37}$$

which is the Fubini-Study metric on $P^n(\mathbf{C})$. Its Kähler form K is

$$K = \frac{\mathrm{i}}{2} \sum_{k=0}^{n} \omega_{0\bar{k}} \wedge \omega_{\bar{0}k} = \frac{1}{2\pi \mathrm{i}} \Omega \tag{38}$$

If $P^1(\mathbf{C})$ is a complex line in $P^n(\mathbf{C})$, integrating equation (38) over $P^1(\mathbf{C})$ gives

$$\int_{P^1(\mathbf{C})} \frac{1}{2\pi \mathrm{i}} \Omega = 1$$

Let

$$\xi = \frac{1}{2\pi \mathrm{i}} \Omega$$

and treat it as a cohomology class in dimension two. Its dual homology class can be viewed as the hyperplane A in $P^n(\mathbf{C})$. Thus, ξ is the first Chern class of the universal bundle.

More generally, if X is any compact Riemann surface and f is a non-constant holomorphic map, f:$X \longrightarrow P^n(\mathbf{C})$, mapping X into $P^n(\mathbf{C})$, then $f(X)$ can be viewed as an algebraic curve in $P^n(\mathbf{C})$. Let E_0 be the universal line bundle on $P^n(\mathbf{C})$, then

$$c_1(f^*(E_0)) = f^*(\xi) = \frac{1}{2\pi i} f^*(\Omega)$$

$$\int_X c_1(f^*(E_0)) = \frac{1}{2\pi i}\int_{f(X)} \Omega = \text{area of } f(X)$$

On the other hand, by the pairing relationship between the upper and lower homology classes, we know

$$\frac{1}{2\pi i}\int_{f(X)} \Omega = \langle f(X), A\rangle = \text{times of } f(X)$$

Thus,

$$\text{times of } f(X) = \frac{1}{2\pi i}\int_{f(X)} \Omega = \text{area of } f(X) \tag{39}$$

In other words, the times of an algebraic curve in $P^n(\mathbf{C})$ is exactly its area.

This perspective has the advantage that it can be extended to Riemann surfaces X with boundaries. In this case, the area of $f(X)$ still makes sense. Let A be a hyperplane in $P^n(\mathbf{C})$, and let $n(f(X) \cap A)$ be the intersection number of $f(X)$ and A.

In this case, $n(f(X) \cap A)$–area of $f(X)$ is no longer zero, and it is actually related to the integral of Ω along the boundary of $f(X)$.

The transgression formula in equation (36) can help us compute this integral. By using equation (36) and Stokes Theorem, this integral can be computed using the connection form $\omega_{0\bar{0}}$.

Since we are considering holomorphic objects, the connection form $\omega_{0\bar{0}}$ can be integrated out, so equation (36) becomes a double transgression formula.

In fact, when $Z \in \mathbf{C}^{n+1} \setminus \{0\}$, let

$$Z_0 = \frac{Z}{|Z|}, \text{ where } |Z|^2 = \langle Z, Z\rangle \tag{40}$$

We can prove

$$\begin{aligned}\omega_{0\bar{0}} &= \langle DZ_0, Z_0\rangle = \langle dZ_0, Z_0\rangle \\ &= \frac{1}{2|Z|^2}\{\langle dZ, Z\rangle - \langle Z, dZ\rangle\} \\ &= (\partial - \bar{\partial})\log|Z|\end{aligned}$$

where $d = \partial + \bar{\partial}$, thus,

$$\Omega = d(\partial - \bar{\partial})\log|Z| = -2\partial\bar{\partial}\log|Z| \tag{41}$$

This double transgression formula is the key to the first fundamental theorem in Nevanlinna theory.

In classical Nevanlinna theory, the Riemann surface X we consider is not compact. If there is a sequence of compact discs D_r of radius r approaching X, the integral on the boundary of X can be approximated by the integral on the boundary of D_r. When we can take a sequence of concentric discs to approximate, the situation simplifies further.

Using equation (41), we can compute the difference between the intersection number of $f(D_r)$ with A and the area of $f(D_r)$.

Let

$$\begin{aligned}&d^c = i(\partial - \bar{\partial}) \\ &n(f(D_r)\cap A) = \text{the intersection number of } f(D_r) \\ &\text{area}(f(D_r)) = \text{the area of } f(D_r) \\ &A^\perp = \text{any non-zero vector in the orthogonal complement of } A \\ &|Z, A^\perp| = |\langle Z, A^\perp\rangle| \\ &|Z| = |\langle Z, Z\rangle|^{\frac{1}{2}} \\ &|A^\perp| = |\langle A^\perp, A^\perp\rangle|^{\frac{1}{2}}\end{aligned}$$

then we can derive a relationship:

$$n(f(D_r)\cap A) - \text{area}(f(D_r)) = \frac{1}{2\pi}\int_{f(\partial D_r)} d^c \log\frac{|Z, A^\perp|}{|Z|\cdot|A^\perp|} \tag{42}$$

We now prepare to integrate equation (42) over r. Define

$$N(r, A) = \int_0^r \frac{n(f(D_t)\cap A)}{t}\,dt$$

$$T(r) = \int_0^r \frac{\text{area}(f(D_t))}{t} \mathrm{d}t$$

and $T(r)$ is called the Nevanlinna characteristic function, which is independent of the choice of A.

Now, it is not difficult to obtain the **first fundamental theorem of value distribution theory**: There exists a constant C such that $N(r,A) \leq T(r) + C$.

This is a very important theorem. It shows that, for non-compact Riemann surfaces, although the intersection number with the hyperplane A is not constant, the function $N(r, A)$, which is related to the number of intersection points, is controlled by $T(r)$, which is independent of the choice of A.

The second fundamental theorem of Nevanlinna theory estimates the lower bound of

$$\sum_{i=1}^{s} N(r,A_i)$$

where A_i is a set of hyperplanes. From our perspective, it applies equation (41) to the case of the canonical bundle. It generalizes the Riemann-Hurwitz and Phragmén-Lindelöf formula.

With the second fundamental theorem, we can consider the **Nevanlinna deficiency formula**:

Let A be a hyperplane. Define

$$\delta(A) = 1 - \overline{\lim_{r \to \infty}} \frac{N(r,A)}{T(r)}$$

which is called the deficiency of the hyperplane A.

Let $C_0 = \{Z \in \mathbf{C} \| Z| < 1\}$, and let $f: C_0 \longrightarrow P^1(\mathbf{C})$ be a non-constant holomorphic map. Let $A_1, A_2, \ldots, A_s$ be distinct points in $P^1(\mathbf{C})$. Then

$$\sum_{i=1}^{s} \delta(A_i) \leq 2$$

Picard Theorem says that except for two values, a non-constant meromorphic function takes all other values. The Nevanlinna deficiency formula clearly encompasses Picard Theorem.

Since Nevanlinna's value distribution theory has recently been treated as a limiting case of algebraic geometry, it has attracted much attention. In fact, by appropriately generalizing its second fundamental theorem to arithmetic geometry, it can unify results like the Roth Theorem in Diophantine approximation theory and the Faltings Theorem in the Mordell Conjecture. Please refer to [5].

A recent fundamental insight in algebraic number theory is the similarity between integers and entire functions. The Diophantine equation, $F(x,y)=0$, for integer solutions is very similar to the same equation with complex coefficients for entire function solutions. (The complex case is much easier!) This relationship will eventually clarify and become a groundbreaking achievement in mathematics.

References

1. Chern S. Complex manifolds without potential theory. 2nd ed. Springer-Verlag; 1979.
2. Chern S. Vector bundles with a connection. In: Global differential geometry, studies in mathematics. Vol. 27. Mathematical Association of America; 1989. 1–26.
3. Ehresmann C. Sur la topologie de certains espaces homogènes. Annals of Math. 1934; 35: 396–443.
4. Steenrod N. The topology of fiber bundles. Princeton University Press; 1951.
5. Vojta Paul. Diophantine approximation and value distribution theory. Lecture Notes in Mathematics. Vol. 1,239. Springer-Verlag; 1987.
6. Witten E. Quantum field theory and the Jones polynomial. In: Yang C. N., Ke M. L., editors. Braid group, knot theory, and statistical mechanics. World Scientific; 1989: 239–329.

What Is Geometry?*

Today's award ceremony is a grand event. Hearing so many wonderful speeches, I feel deeply touched. It is truly an honor and a great pleasure to have the opportunity to speak here today. Let's approach this session as if it were a regular class—informal and relaxed. I've brought along some materials, but unfortunately, I cannot project them. That's okay, as I didn't prepare much in advance anyway.

I am expected to talk about geometry, titled "What Is Geometry." Although I have worked in the field of geometry for several decades, I must apologize that when you leave here, you might not walk away with a simple answer. Geometry is a vast and profound discipline, one that has seen monumental development over thousands of years. It is not only closely intertwined with many other sciences, but it also plays a foundational role in them.

When we talk about geometry, the first name that naturally comes to mind is Euclid. Apart from the Christian Bible, Euclid's *Elements* is arguably the most widely circulated book in history. This book was translated into Chinese by Xu Guangqi and Matteo Ricci. Xu Guangqi (1562–1633) was a distinguished Chinese scholar, while Matteo Ricci was an Italian missionary to China. Together, they translated only the first six chapters, with the Chinese edition published in 1607. Many of the geometric terms we use today—such as parallel lines, triangles, and circumference—were likely coined by Xu Guangqi.

They didn't complete the entire translation. About half the book was translated by another duo: Li Shanlan and Alexander Wylie. Wylie was a British missionary, while Li Shanlan hailed from Haining in Zhejiang Province. Haining is a county in Jiaxing Prefecture, and since I'm from Jiaxing, we share the same hometown. (Applause) In fact, Mr. Zha Jiming is also from Haining. (Applause)

---

* This article is the academic lecture delivered by Professor Chern on September 24, 1999, during the prestigious Qiushi Foundation Science Prize award ceremony at Fudan University. Notably, it also served as the inaugural lecture of the esteemed Yang Wuzhi Lecture Series at Fudan University.

The second figure who profoundly influenced the history of geometry is René Descartes (1596–1650). Descartes was a French philosopher rather than a dedicated mathematician. By introducing the coordinate system, he revolutionized geometry by transforming it into algebra. At that time, there was no concept of analysis or infinity, so he became algebra. I guess Descartes might not have realized how revolutionary his work was. His essay on geometry was simply an appendix to his philosophical writings.

Back then, this idea was transformative for geometry. By expressing geometric phenomena in terms of coordinates, he turned them into algebraic problems. For instance, to prove that a line passes through a specific point, you only need to show that a certain value equals zero—a relatively simple algebraic problem. Of course, not every geometric problem becomes an algebraic one; in some cases, it becomes even more complex. Still, the relationship between geometry and algebra is fundamental. Most of us were introduced to Descartes' coordinate system in high school analytic geometry.

One noteworthy detail—though I regret being unable to project my diagram. He began with a straight line, designated an origin on it, and defined points by their distance, x, from the origin. Another line was drawn in a specific direction from the origin, and the y-coordinate was determined by the distance of a point from the x-axis.

This is the Cartesian coordinate system, named after Descartes. This coordinate system does not necessarily require perpendicular axes. It's unclear which scholar, when writing textbooks, depicted the two lines as perpendicular, thus making the x-coordinate and y-coordinate symmetrical. Descartes's two axes were not symmetrical—a critical concept that we now call a fiber bundle. Lines parallel to the y-axis represent fibers, which form another space. And here is why: By changing it, the straight line no longer needs to be a straight line—it could represent any space. In this way, the points in a space can be defined using entirely new sets of coordinates. So sometimes science or mathematics doesn't always progress; at times, it may even regress. (Laughter)

Using his coordinate system, Descartes also discovered that we are not limited to Cartesian coordinates. For example, we can use polar coordinates. To define a point on a plane, we designate a point as the origin and draw a ray called the polar axis. Each point on the plane is then defined by its distance from the origin and the angle between this ray and the line connecting the point to the origin.

In polar coordinates, one value is a positive number or zero, and the other is an angle between 0° and 360°. Of course, we all know that there are many other coordinate systems—as long as a set of numbers can define a point, we call it a coordinate system.

Over time, mathematicians made another revolutionary breakthrough—coordinates don't necessarily need to have inherent meaning. As long as a set of numbers defines a point, it qualifies as a coordinate system. Consequently, geometric properties became algebraic or analytical properties of coordinates. This quantification of geometry turned it into something formalized, with profound implications.

Yet, these ideas were not immediately embraced. Even Einstein struggled with them. He developed his special theory of relativity in 1908 and his general theory of relativity in 1915—a seven-year gap. Einstein explained that it took him so long to transition from special to general relativity because he believed coordinates must have geometric or physical meaning.

Einstein, as a meticulous scholar, found it difficult to accept meaningless coordinates, which delayed his progress for many years. He had no choice but to accept it because the concept of space had been expanded.

I left out a part earlier. It doesn't matter if I miss something while discussing a book—it's not a big deal if I forget, and it's perfectly fine even if I make a mistake. (Laughter) Let me go back and share a little more about Euclid. Back then, Euclid's *Elements* was not merely a work of geometry, but the entire mathematics. This was a time before the discovery of calculus, and although the concept of "infinity" existed, it hadn't been widely embraced. Additionally, it's unfortunate that we know very little about Euclid's personal background. We know only that he lived around 300 BC and was a geometry professor at the University of Alexandria. His *Elements* was likely a textbook of its time. Alexandria was the final hub of Greek culture, and because Alexander the Great had visited, he established a major city in North Africa along the Mediterranean Sea. However, we know, after Alexander's campaign in Asia, he passed away shortly after. His general, Ptolemy Soter, took control of Egypt. Ptolemy highly valued education and founded a university next to his palace. It was the world's greatest university at that time, with superb facilities and an incredible collection of books. Sadly, due to religious opposition and other factors, the university was completely destroyed. Early Christianity didn't appreciate the institution, and it began to be dismantled. Later, when Muslims took control of

North Africa, they inflicted widespread destruction, burning the books in the library. So, unfortunately, that great university no longer exists.

Geometry is of immense importance, as it was once considered synonymous with mathematics itself. For example, the French Academy of Sciences still refers to its mathematics division as the Geometry Group. In France, mathematicians were often called geometrists rather than mathematicians. At that time, geometry had a much broader scope than it does today.

However, today, the scope has expanded even further—just as I mentioned earlier, coordinates don't necessarily have inherent meaning. A space can have multiple coordinate systems. But how do we describe space? This appears difficult: What are the true geometric properties of space? This is a major question, and around it Gauss and Riemann advanced the theory significantly. Gauss, a German, in my opinion, was the greatest modern mathematician, and Riemann, his successor, was also a German mathematician. Both were professors at the University of Göttingen. Unfortunately, Riemann suffered from poor health and tuberculosis and passed away at the age of 40. Their main goal was to develop a theory of space that utilized local coordinates. In such a space, there are only coordinates, which cannot be explicitly defined; they are simply markers for points. While the properties of these points can be expressed through analytical relationships, studying space itself becomes a big challenge.

Before this, I forgot another as well: The foundation of mathematics lies in Euclid's book; however, there's an issue with his book. He used axioms and logical reasoning to derive conclusions. For instance, consider the famous result that the sum of the interior angles of a triangle is always 180°. This is a remarkable finding, and Euclid was able to prove it in just a few steps using his axioms. It is a conclusion and of greater simplicity than modern science. We have heard a lot that when scientists conduct research, the first step is often to request funding from governments or society, stating, "With funding, I can perform experiments."

Of course, experimentation forms the foundation of science. However, this approach inevitably introduces social and political challenges. Euclid said, "Give me a piece of paper, and I'll write down a proof." Moreover, as a mathematician, I believe mathematics has another unique advantage—it can predict experimental results. Without experiments, we can derive conclusions via mathematics and then use experiments to verify them. Of course, there are cases where experiments fail to confirm a theory, suggesting flaws in the theory itself, but that's a separate matter.

Euclid's axioms are generally intuitive, but his famous fifth postulate caused considerable debate. This fifth postulate, though somewhat verbose, can be summarized as follows: "Given a line and a point not on that line, there is exactly one line through the point that is parallel to the given line." If you casually draw this, it seems quite believable. However, upon closer scrutiny, this axiom is not as self-evident as the others. For mathematically inclined thinkers of the time, this postulate became a significant problem. The ideal scenario would have been to derive the fifth postulate from the other axioms, turning it into a theorem. This would simplify the problem and make the experiment feasible.

Mathematicians often employ a straightforward method for proving such claims—they assume the statement is false and check whether it leads to a contradiction. If a contradiction arises, the statement must be true. This approach, known as the indirect proof method, was applied to the fifth postulate, but all attempts failed. We now know that the fifth postulate is not universally true—through a given point, there can be infinitely many parallel lines. This realization marked the discovery of non-Euclidean geometry.

The discovery of non-Euclidean geometry had profound social implications because it suggested that space might not be unique. In Western societies, where monotheism prevailed, the idea of multiple or different kinds of spaces was difficult to reconcile. At the time, this was a serious social issue, as well as a philosophical one. Figures like the German philosopher Immanuel Kant believed that only Euclidean geometry was possible, dismissing the notion of non-Euclidean spaces. This sparked intense debates.

The formal discovery of non-Euclidean geometry is credited to János Bolyai, a Hungarian, in 1832, and Nikolai Lobachevsky, a Russian, in 1847. However, the renowned mathematician Carl Friedrich Gauss had also understood it, as evidenced by his works. Yet, Gauss chose not to publish his findings, likely fearing criticism and controversy. Thus, the debate over the fifth postulate persisted.

Later, the Italian geometer Beltrami, by constructing a surface within the framework of Euclidean three-dimensional space, showed that the geometry on this surface is non-Euclidean. This provided a powerful means to dispel the skepticism surrounding non-Euclidean geometry. His work demonstrated that, within the context of three-dimensional Euclidean space, it is possible to create a non-Euclidean space. Hence, non-Euclidean geometry can also be present within Euclidean geometry. In essence, if you accept Euclidean geometry, you must also acknowledge the validity of non-Euclidean geometry. This realization gradually helped to dissolve the doubts and resistance toward non-Euclidean geometry.

Earlier, I mentioned how Gauss and Riemann generalized the concept of coordinates, transforming them into abstract constructs that no longer required direct, intuitive meaning. This innovation profoundly impacted geometry, as it redefined space into a patchwork of segments, pieced together in complex ways. So how do we study such spaces? How can we understand their unique properties? And how can we differentiate one type of space from another?

I wish I could show you some diagrams I drew that illustrate different kinds of spaces, but unfortunately, I cannot project them here. Nevertheless, the point is that the number of possible spaces is infinite. There exist countless varieties, each with unique characteristics. This realization gave rise to a pivotal question for geometers—how should we approach the study of such spaces? From this emerged an entirely new field of mathematics: topology.

Topology focuses on studying the intrinsic properties of spaces. It delves into questions: What defines the continuity of a space? Under what conditions can two spaces be considered the same in some meaningful sense? Such inquiries have led to the development of a vast array of mathematical tools.

Riemann also grappled with these profound questions. Born in 1826 and passing in 1866, Riemann contributed during an era when the German education system required aspiring professors to deliver a public lecture—known as a Habilitationsschrift—to qualify for university teaching positions. In 1854, at the University of Göttingen, Riemann gave such a lecture upon becoming a professor. This lecture is one of the foundational documents in modern geometry, addressing such questions.

How to study such spaces? To do so, if one only knows a space is merely an abstract collection of patches, there's little to study. Researchers thus often introduce a metric, a way of defining the distance between two points within the space. But how does one measure such distances? Analytical tools become indispensable. Distances are frequently expressed as integrals, with the integral representing the distance between two points.

Riemann's 1854 paper is of immense significance, and it stands as one of the foundational works in geometry, much like the constitution of a nation. Einstein, unaware of this paper, spent seven years struggling to develop the same concepts, wasting a considerable amount of time. The notion of distance introduced by Riemann in this paper—expressed as an integral—has undergone tremendous development in the mathematical community over the past century. The first major application was the use of Riemannian geometry in general relativity, forming the mathematical foundation of Einstein's theory of relativity. Today, anyone studying mathematics, particularly geometry, must delve into Riemannian geometry, which has become a central part of the field. And that started with this lecture of his. The results derived from Riemannian geometry are countless, and it is not only the foundation of geometry but perhaps also the very cornerstone of the development of mathematics as a whole.

I just mentioned the development over the past hundred years. The so-called Riemannian geometry is actually a simple case of Riemann's paper, a particular case. The original meaning of Riemann, in a broader sense, was greatly influenced by the work of a German mathematician, Finsler. Therefore, this part of geometry is called Finsler geometry. In 1918, he wrote a doctoral dissertation at the University of Göttingen, discussing this geometry. This geometry didn't develop much afterward because people didn't know what to do with it. If the integral of this metric is a bit broader, the corresponding mathematics becomes more complicated, unlike Riemann's specific case, which was simple. Even Riemann's case was not simple. Riemann simply wrote that the square of ds equals a second-order differential expression, and integrating this differential expression gives the length of an arc. How to study this kind of geometry requires a genius like Riemann, who developed what he called the Riemann curvature tensor. If you want to study this kind of geometry, you must have the concept of tensors. The curvature of space, when represented analytically, becomes quite complicated, which is the Riemann curvature tensor.

Today, people like to talk about awards. When we give awards, there is often a monetary reward, and society and the government show respect for your work. Back then, if you wanted to do mathematics but didn't have a position as a mathematics professor, no one would pay you a salary. One main way to survive was to win awards. Several academies of science gave awards, and after winning one, you could sustain yourself for a while, which was quite satisfying. However, what's interesting is that I believe Riemann's discovery of the Riemann-Christoffel curvature tensor was a great achievement. He went to the French Academy of Sciences to apply for an award, but the members of the Academy couldn't understand his work, so they didn't give him the award.

So, to those of you sitting in the front rows, you are all award winners, and you are honored. We have high expectations for you. However, most people in the back rows haven't won awards. But I want to comfort everyone that it doesn't matter if you haven't won an award—you can still do great work. I, for example, didn't win any awards before I got my degree. Winning an award is not the most important factor. Riemann didn't win an award either. His Riemann-Christoffel tensor was not recognized by the French Academy of Sciences when he applied for an award.

Recently, although there has been remarkable development in Riemannian geometry, there has not been much progress in the general case, the general case in Riemann's paper—Finsler geometry. Interestingly, in 1942, I wrote a paper on Finsler geometry, where I extended Riemann's results to the Finsler case. Recently, two young Chinese mathematicians, Bao Dawei and Shen Zhongmin, and I coauthored a book on Finsler geometry. This book is about to be published by Springer as part of their *Graduate Texts in Mathematics* series. The editor of the series really liked our book and gave it an interesting book number: 2000. The book is here, and I'll later give it to Professor Gu Chaohao to be placed in the library at Fudan University. (Applause) Our book has a small achievement: We have extended the recent discoveries in Riemannian geometry over the past century to the general case, namely the Riemann-Finsler case. This was Riemann's original intention. Riemann was of course very great, but he didn't emphasize the general case. He even said in his paper that there was nothing new in it. Well, we've taken what he referred to as nothing new and made some progress on it.

I know that I am sitting beside two great physicists. Now, I'd like to take the liberty of discussing the relationship between physics and geometry. In my view, many significant developments in physics stem from the idea that physics is, at its core, geometry.

Consider Newton's second law of motion, $F = ma$. Here, F represents force, m is mass, and a is acceleration. But acceleration, in modern terms, is called curvature. This means that the right-hand side of the equation is a geometric quantity, whereas force is a physical quantity. So after all his efforts, Newton was essentially asserting that physics is geometry! (Laughter and applause)

Moreover, this idea extends beyond Newton—Einstein's general theory of relativity follows the same principle. The Einstein field equations are given by

$$R_{ik} - \frac{1}{2} g_{ik} R = 8\pi K T_{ik}$$

Here, R_{ik} is the Ricci curvature, R is the scalar curvature, K is a constant, and T_{ik} is the energy-stress tensor. If you examine this equation closely, you'll see that the left-hand side consists entirely of geometric quantities derived from the Riemannian metric. In other words, Einstein's fundamental equation states that geometric quantities are equal to physical quantities! (Applause)

But this idea doesn't stop here—we can keep going. Today, the spaces we study in mathematics are called manifolds, which can be thought of as pieces of space stitched together. These manifolds are challenging to analyze. To describe the metric of a manifold using equations, we must first linearize it, which means introducing a vector space—a mathematical structure that allows for vector addition, subtraction, and various forms of multiplication. This, in turn, enables us to apply analytic methods to geometry.

How do we handle a general manifold? The mathematical approach is straightforward: At every point on the manifold, we attach a tangent plane. Each of these tangent planes forms a vector space known as the tangent space. In Euclidean geometry, there is only one tangent space. But as space is more complicated, every point has its own tangent space, all of which remain flat. This leads to a profound development in geometry: Instead of limiting ourselves to tangent spaces, we can assign a vector space to each point of a manifold—without necessarily making it the tangent plane or space. This broader structure is called a fiber bundle, also known as a vector bundle.

I would argue that this concept is even more important than Einstein's theory of relativity. Maxwell's equations are built upon a vector bundle. Instead of just assigning a collection of vector spaces, the best analogy is a bundle of chopsticks—except that in this case, each chopstick represents a one-dimensional complex space. Though we call it one-dimensional, it is actually two-dimensional in the complex sense. And the moment you introduce complex numbers, a world of possibilities opens up.

Now, with complex lines, if we can establish a notion of parallel transport between fibers, we immediately arrive at Maxwell's equations. Modern civilization depends on electricity, and the fundamental equations governing electricity are Maxwell's equations. In the language of fiber bundles, the differential of parallel transport gives the field strength F, and applying another differential operation (the codifferential) to F yields the current vector J. Using just two simple equations, we can express Maxwell's equations as: $\mathrm{d}A = F, \delta F = J$.

Normally, if you read a textbook on electromagnetism, you need to understand the meaning of electromagnetism. I don't understand it. However, when these textbooks lay out the equations, they often take up an entire page, filled with various differential terms and lengthy explanations.

But in simple terms, Maxwell's equations can be understood as follows: The differential of parallel transport gives the field strength, and through a certain operation on the field strength, we obtain the current vector. This is precisely Maxwell's equations, expressed in a form that is entirely equivalent to the original.

Thus, Maxwell's equations are nothing more than the physics of a one-dimensional fiber bundle—albeit a complex one-dimensional bundle. But how do we stitch these fibers together? For that, we need the concept of a group. A group provides a transformation rule that allows one fiber to be moved to another. If the fiber bundle is one-dimensional, even in the complex case, we only need an Abelian (commutative) group.

This is where Professor Chen-Ning Yang made a groundbreaking contribution. He extended this concept to a non-Abelian group, specifically the SU(2) group. Using SU(2) connections, we obtain a parallel formulation of Maxwell's equations, but now generalized to what we call the Yang-Mills equations: $DA = F, \delta F = J$.

This has tremendous significance. For those of us working in geometry, discovering such connections is already remarkable. But for physicists to acknowledge that these relationships have physical meaning is even more difficult—and fundamentally important.

Take, for example, last year's Nobel Prize winner, whom I'm sure you all know—Daniel Tsui—who worked on the theoretical aspects of the so-called Hall effect. His research also utilized some of our work. We focus on curvature. If you're driving a car, the more curved the road, the slower you need to go; if the road is straight, you can speed up—this is curvature. In higher dimensions, curvature becomes more complex, but it can still be elegantly handled using algebra.

One of my friends, also my student, Simons, worked on curvature-related research, and our contributions have benefited Tsui and his team in their Nobel-winning work. Generally speaking, in mathematics, we see ourselves as caretakers—sweeping the floor, tidying things up, and clarifying structures. Then, when great physicists come along and tell us that our work actually makes sense in their domain—well, that makes us very happy. (Laughter and applause)

Today, geometry is not only applied in physics but also in biology. Take the structure of DNA, for example—it's a double helix, and this structure is deeply tied to geometry. Many geometric studies are dedicated to understanding this. In fact, at major universities now, students studying biology are required to learn geometry.

Now, researchers are investigating larger molecular compounds, which are essentially formed by the arrangement of atoms. The way these atoms are arranged is a geometric problem. These geometric concepts are no longer abstract; they have real chemical significance.

One advantage mathematics has over other sciences is that it fundamentally remains an individual pursuit. Even in remote places, progress is possible. Of course, it helps to have a few friends with whom one can exchange ideas.

Thank you, everyone. (Thunderous applause)

Gauss-Bonnet Theorem and Maxwell's Equations*

Elementary Geometry

One of the most fundamental special cases of the Gauss-Bonnet Theorem is the well-known Triangle Angle Sum Theorem: In Euclidean geometry, the sum of the interior angles of a triangle is equal to π.

This theorem can be proved as follows:

Consider a triangle ABC in the Euclidean plane. Draw a straight line XAY through vertex A, parallel to the opposite side BC. Since alternate interior angles are equal, we have $\angle B = \angle XAB$, $\angle C = \angle YAC$. Therefore, the sum of the interior angles is $\angle A + \angle B + \angle C = \angle XAB + \angle BAC + \angle CAY = \pi$.

Thus, the theorem is proved.

Alternatively, using exterior angles, we can restate the theorem as follows: The sum of the exterior angles of a triangle is 2π.

The advantage of this formulation is that it generalizes to any convex polygon: For any convex polygon in the plane, the sum of its exterior angles is always 2π.

---

* This article is a popular academic lecture given by Professor Shiing-Shen Chern to students at the Department of Mathematics, Nankai University. It was originally published in *Science* 53, no. 3 (2001).

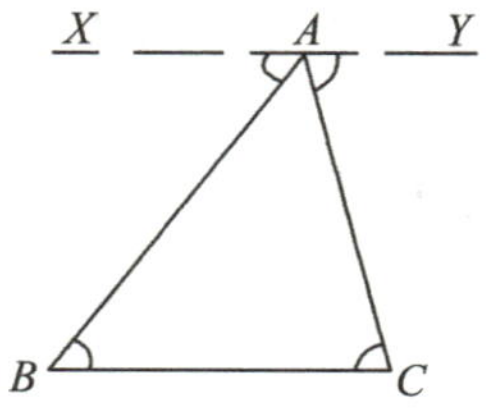

(a) The sum of the interior angles of a triangle is equal to π.

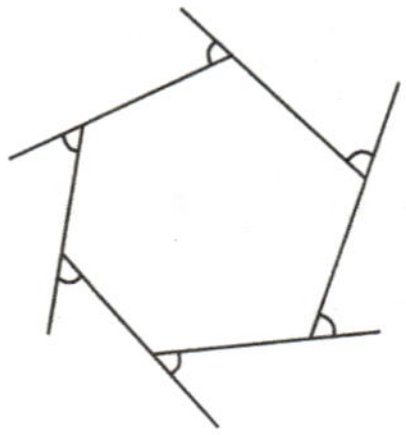

(b) The sum of the exterior angles of any convex polygon is equal to 2π.

Fig. 1 Proof of the polygon interior angle sum theorem

This concept can be extended further. Consider any simple closed curve C (a curve that does not intersect itself). Take a point P on C and let PT be the unit tangent vector at P. At a fixed reference point O, construct a unit vector OQ parallel to PT.

According to "tangent rotation" theorem, as P moves along C and completes one full traversal, the vector OQ rotates by a total angle of 2π.

This theorem requires proof.[1] Note that the curve C can be quite curved, causing Q to move forward and backward on the unit circle. The theorem states that after canceling out these movements, the point Q completes exactly one full rotation.

By combining these two results, the theorem can be extended to smooth curves with cusps.

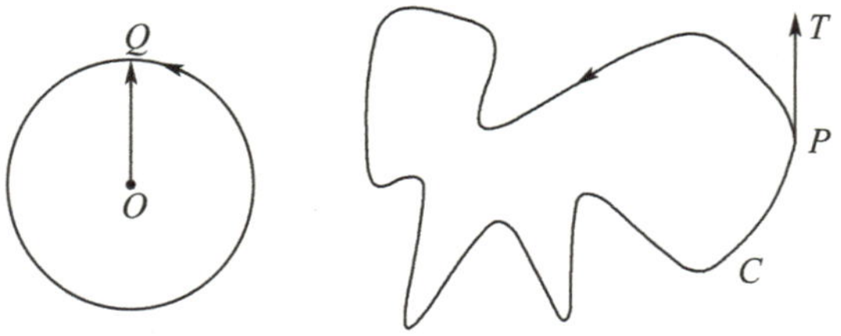

As P moves along any simple closed curve C, the unit vector OQ, which is parallel to the tangent PT at P, rotates by an angle of 2π.

Fig. 2 "Tangent rotation" theorem

This theorem can also be extended to self-intersecting curves, making it relevant to homotopy theory in topology.

Now, let's consider the same problem on a sphere. If we regard great circles on a sphere as straight lines, the geometry becomes non-Euclidean, commonly known as elliptic geometry.

Now, let's examine a spherical triangle. Let r be the radius of the sphere. Consider a spherical triangle ABC formed by great circle arcs. Extending the two sides AB and AC beyond vertex A, they intersect at the antipodal point A^*. The two great circles ABA^* and ACA^* form a spherical sector, whose area is given by $(A/2\pi)\cdot 4\pi r^2 = 2r^2 A$, where A is measured in radians. Performing the same construction at the other two vertices B and C, we obtain three such spherical sectors.

Now, by merging all antipodal points of the sphere, we obtain an elliptic plane in non-Euclidean geometry. It is evident that the three sectors together cover the entire plane, while the triangle ABC itself is counted three times in this covering process. Let Δ denote the area of the spherical triangle. The total covered area is $2\pi r^2 + 2\Delta$. The first term represents the area of the elliptic plane, i.e., half of the sphere's surface area $= 2\pi r^2$.

Thus, we derive the equation

$$A + B + C = \pi + (1/r^2)\Delta \qquad (1)$$

This formula states that the excess angle sum, given by $A + B + C - \pi$, is proportional to the triangle's area, with a factor of $1/r^2$. As a result, in spherical geometry, the sum of the angles of a triangle is always greater than π.

In hyperbolic non-Euclidean geometry, the situation is reversed—the sum of a triangle's angles is always less than π.

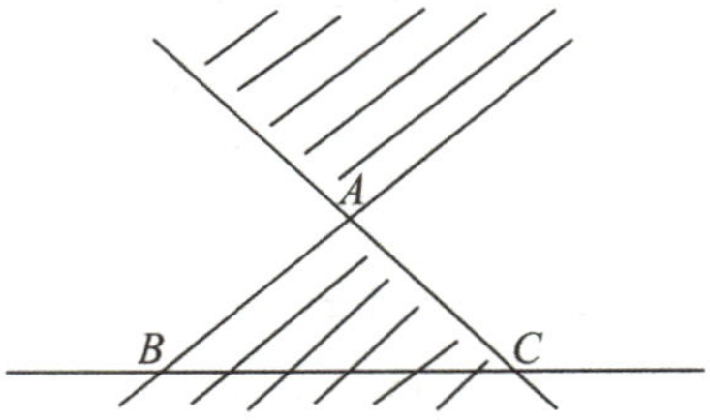

Fig. 3 Proof that the excess angle sum $A + B + C - \pi$ is equal to $1/r^2$ times the area

Gauss-Bonnet Theorem on Surfaces

The previous results can be extended to triangles on surfaces, known as the Gauss-Bonnet formula.

Let S be a surface, C a piecewise smooth closed curve on S, and D the region enclosed by C. Let k be the geodesic curvature of C, and K the Gaussian curvature at each point of D. Let A represent the angle at non-smooth points. The Gauss-Bonnet formula is given by

$$\sum(\pi - A) + \int_C k\mathrm{d}s + \iint_D K\mathrm{d}\sigma = 2\pi\chi \qquad (2)$$

where $\mathrm{d}\sigma$ is the surface area element, and χ is the Euler characteristic of the region D.

In the case of the first section, C is a geodesic, so $k = 0$. In the case of a Euclidean triangle, $K = 0$, and $\chi = 1$, which gives $\sum A = \pi$. In the case of spherical geometry, where the radius is r, we have $k = 1/r^2$, and in the formula (2), the integral along C is 0, and $\chi = 1$. Thus, from formula (2), we recover equation (1).

Formula (2) is one of the most important theorems in surface theory, and its proof can be found in the reference [2]. Interestingly, this formula is not often found in standard textbooks. On the left-hand side of formula (2), the terms represent curvatures: The first term is the point curvature, the second term the line curvature, and the third term the surface curvature. Formula (2) expresses total curvature as a topological invariant.

Around 1940, when I was teaching elementary differential geometry at the National Southwest Associated University in Kunming, I provided proof of this formula. This result can also be generalized to higher dimensions, becoming one of my later significant contributions, briefly outlined as follows:

This proof is based on the concept of connection, utilizing exterior derivatives.

Let x be a point on the surface S, and let e_1 be a unit tangent vector at x. Assume that S is oriented. Then, the unit tangent vector e_2, which is perpendicular to e_1 at x, is completely determined by the orientation. The set of all unit tangent vectors e_1 forms a space, which is three-dimensional, because the trajectory of the point x on the surface S is two-dimensional, and for the same point x, e_1 forms a circle. This space is called E. By taking the origin of the unit tangent vectors, we obtain the map

$\pi: E \to S$. This structure is called a circle bundle, a special type of fiber bundle. For $x \in S, \pi^{-1}(x)$ is the circle of unit tangent vectors at x.

An important quantity is

$$\omega = \mathrm{d}e_1 \cdot e_2 \tag{3}$$

Here, $\mathrm{d}e_1$ represents the differential of the vector e_1, and the product is the inner product of the vectors. Thus, ω is a one-form on the space E, known as the connection form. Its exterior derivative is

$$\mathrm{d}\omega = -\pi^*(K\mathrm{d}\sigma) \tag{4}$$

This is a remarkable formula that Gauss would certainly have appreciated.

The Gauss-Bonnet formula follows naturally by integrating formula (4). The proof of formula (4) can be found in the appendix.

Maxwell's Equations

The above discussion does not have to be limited to the tangent bundle of a surface. We can take any family of Euclidean two-dimensional planes, whose parameter space is an arbitrary manifold. In this case, there is also a connection form ω, and its exterior derivative is given by

$$\mathrm{d}\omega = \Omega \tag{5}$$

which is a two-form on the base space S, called the curvature form. Since $\mathrm{dd} = 0$, we have

$$\mathrm{d}\Omega = 0 \tag{6}$$

In the case when S is a four-dimensional Lorentzian manifold, we obtain Maxwell's equations. Let $x_\alpha (0 \leq \alpha \leq 3)$ be the coordinates of S, and the Lorentz metric is

$$\mathrm{d}s^2 = -\mathrm{d}x_0^2 + \mathrm{d}x_1^2 + \mathrm{d}x_2^2 + \mathrm{d}x_3^2 \tag{7}$$

Define

$$\Omega = \sum_{\alpha,\beta} F_{\alpha\beta} \mathrm{d}x_\alpha \wedge \mathrm{d}x_\beta, F_{\alpha\beta} = -F_{\beta\alpha} \tag{8}$$

where

$$(F_{\alpha\beta}) = \frac{1}{2}\begin{pmatrix} 0 & E_1 & E_2 & E_3 \\ -E_1 & 0 & -B_3 & B_2 \\ -E_2 & B_3 & 0 & -B_1 \\ -E_3 & -B_2 & B_1 & 0 \end{pmatrix} \tag{9}$$

For the metric $\mathrm{d}s^2$, there is the * operator, defined as

$$\mathrm{d}^* = * \mathrm{d} * \tag{10}$$

Define the current $j = (j_1, j_2, j_3)$, and

$$J = -\rho \mathrm{d}x_1 \wedge \mathrm{d}x_2 \wedge \mathrm{d}x_3 + \mathrm{d}x_0 \wedge (j_1 \mathrm{d}x_2 \wedge \mathrm{d}x_3 + j_2 \mathrm{d}x_3 \wedge \mathrm{d}x_1 + j_3 \mathrm{d}x_1 \wedge \mathrm{d}x_2) \tag{11}$$

Then, the equation

$$\mathrm{d}\Omega = 0, \mathrm{d}^*\Omega = 4\pi J \tag{12}$$

can be written as a system of equations:

$$\begin{gathered} \operatorname{div} B = 0, \operatorname{curl} E + \frac{\partial}{\partial t} B = 0 \\ \operatorname{div} E = 4\pi\rho, \operatorname{curl} B - \frac{\partial}{\partial t} E = 4\pi j \end{gathered} \tag{13}$$

Here, $t = x_0$. If t is time, $E = (E_1, E_2, E_3)$ is the electric field vector, $B = (B_1, B_2, B_3)$ is the magnetic field vector, ρ is the charge density, and j is the current density. These are precisely Maxwell's equations.

In J, the components $\rho \mathrm{d}x_1 \wedge \mathrm{d}x_2 \wedge \mathrm{d}x_3$ represent the charge,

$$(j_1 \mathrm{d}x_2 \wedge \mathrm{d}x_3 + j_2 \mathrm{d}x_3 \wedge \mathrm{d}x_1 + j_3 \mathrm{d}x_1 \wedge \mathrm{d}x_2)$$

represent the electric flux $(j \cdot \mathrm{d}S)$, and $\mathrm{d}x_0 \wedge (j \cdot \mathrm{d}S)$ represents the electric current passing through the surface $\mathrm{d}S$ over time: $\mathrm{d}t = \mathrm{d}x_0$.

Interestingly, from $\mathrm{dd} = 0$, we immediately obtain $\mathrm{d}J = 0$, which is precisely the law of conservation of charge:

$$\frac{\partial \rho}{\partial t} + \operatorname{div} j = 0$$

Thus, the geometric foundation of Maxwell's equations is a Euclidean plane bundle, with the base space being a 4-dimensional Lorentzian manifold.

The introduction of the concept of the plane bundle is credited to Weyl, and this is the first gauge theory in physics. At the time, Einstein had a lengthy opposition to it. The physical difficulty arises because the gauge variables are real numbers. If we use periodic variables or complex numbers, everything becomes consistent. In electromagnetism, we use circle bundles or complex line bundles, which aligns with the Gauss-Bonnet formula. Complex numbers simplify mathematics (which is why every algebraic equation has a solution), and they also rationalize physics.

Complex Vector Bundles

The Euclidean $x-y$ plane can be expressed using complex coordinates $z=x+\mathrm{i}y$ making it a complex line. Its Euclidean metric then transforms into a Hermitian metric. Therefore, the mathematical foundation of electromagnetism is a Hermitian line bundle over a Lorentzian manifold.

This concept can be extended to higher-dimensional complex vector bundles and Hermitian space bundles. When working with complex spaces, the appropriate gauge group is the unitary group $U(n)$ or, for simplicity, its subgroup $SU(n)$, consisting of matrices with determinant 1.

Here, Chen-Ning Yang made a groundbreaking contribution: This two-dimensional complex vector bundle has a fundamental physical meaning. The Maxwell's equations above can be generalized, leading to the Yang-Mills equations. Einstein, in his later years, struggled to find a unified field theory and experimented with various geometric spaces. We now know that a single space is insufficient—fiber bundles are required.

From Newton to Maxwell and then to Yang, theoretical physics has progressed along a grand path.

Appendix

From a mathematical perspective, the foundation is still the theory of surfaces in Euclidean space. Here is a brief introduction:

Historically, the most valuable reference is Gauss's theory of surfaces. Dab has written four volumes on this topic, which are fascinating to read and can be considered treasures of human thought.

When studying Euclidean space, we first need to understand its motion group. Let the coordinates of the space be x_1, x_2, x_3. The Euclidean motion is defined by the following transformation:

$$x_i' = \sum_k a_{ik} x_k + b_i, 1 \le i \le 3 \tag{14}$$

where $A = (a_{ik})$ is an orthogonal matrix. We have $\det A = \pm 1$.

We will restrict ourselves to the case where $\det A = +1$, as it does not change the orientation of space.

The motion (14) induces a homogeneous transformation for the vector $v = (v_1, v_2, v_3)$:

$$v_i' = \sum_k a_{ik} v_k, 1 \le i \le 3 \tag{15}$$

For $w = (w_1, w_2, w_3)$, the inner product is defined as

$$\upsilon \cdot \omega = \sum_k \omega_k \upsilon_k \tag{16}$$

The inner product remains invariant under the transformation.

A Euclidean reference frame consists of a point x and three mutually perpendicular unit vectors $e_i (i = 1,2,3)$ passing through x, such that $e_1 e_2 e_3$ align with the orientation of space. The set of all such $e_1 e_2 e_3$ forms a six-dimensional manifold, known as the Euclidean motion manifold. There exists exactly one motion that maps one reference frame to another. Define

$$\begin{aligned} \mathrm{d}x &= \sum_i \omega_i e_i \\ & \qquad\qquad 1 \le i \le 3 \\ \mathrm{d}e_i &= \sum_j \omega_{ij} e_j \end{aligned} \tag{17}$$

Using $e_i \cdot e_j = \delta_{ij}$, we obtain

$$\omega_{ij} + \omega_{ji} = 0 \tag{18}$$

Note that dd = 0, and thus

$$\begin{aligned} \mathrm{d}\omega_i &= \sum_j \omega_j \wedge \omega_{ji} \\ \mathrm{d}\omega_{ij} &= \sum_k \omega_{ik} \wedge \omega_{kj} \end{aligned} \tag{19}$$

This is the Maurer-Cartan equation, where ω_i and ω_{ij} are called the Maurer-Cartan forms.

Let S be a surface, and take a given orientation for S, with e_3 as the corresponding unit normal vector. Then we have

$$\omega_3 = 0 \tag{20}$$

From this, we get $\mathrm{d}\omega_3 = \omega_1 \wedge \omega_{13} + \omega_2 \wedge \omega_{23} = 0$. Thus,

$$\omega_{13} = a\omega_1 + b\omega_2, \omega_{23} = b\omega_1 + c\omega_2 \tag{21}$$

The surface has two fundamental forms:

$$\begin{aligned} I &= \mathrm{d}x \cdot \mathrm{d}x = \omega_1^2 + \omega_2^2 \\ II &= -\mathrm{d}x \cdot \mathrm{d}e_3 = a\omega_1^2 + 2b\omega_1\omega_2 + c\omega_2^2 \end{aligned} \tag{22}$$

The eigenvalues of II with respect to I are called the principal curvatures. Their elementary symmetric functions are

$$H = \frac{1}{2}(a+b),\ K = ac - b^2 \tag{23}$$

They are respectively called the mean curvature and the Gaussian curvature. These two curvatures have a significant role in determining the shape and geometric properties of the surface.

From equation (19), we have

$$d\omega_{12} = -\omega_{13} \wedge \omega_{23} = -K\omega_1 \wedge \omega_2 \tag{24}$$

Note that $\omega_{12} = \omega$, $\omega_1 \wedge \omega_2 = d\sigma$. Thus, we obtain the proof of equation (4).

Equation (24) encompasses Gauss's Theorema Egregium (remarkable theorem, the integral form that leads to the proof of the Gauss-Bonnet formula), which can be considered a treasure of mathematics.

References

1. Klingenberg W. A course in differential geometry. GTM 51. New York: Springer-Verlag; 1978.
2. Blaschke W. Vorlesungen über Differential Geometrie. Bd1. Berlin: Springer; 1944.

Appendix

Mathematical Awards Received by Shiing-Shen Chern

August 1983: Shiing-Shen Chern was awarded the Steele Prize by the American Mathematical Society for the "long-term influence of his entire mathematical work." The award citation recognized him as "The leader in the field of differential geometry for half a century. His work is both profound and elegant, a typical example being his intrinsic proof of the Gauss-Bonnet formula."

May 1984: Chern received the Wolf Prize for the 1983–1984 academic year in recognition of "his profound contributions to global differential geometry, which have influenced all of mathematics."

December 1, 2002: Kazan University in Russia announced that Chern was awarded the Lobachevsky Prize (3rd edition) for his "outstanding contributions to the field of geometry."

May 27, 2004: Chern was honored with the inaugural Shaw Prize in Mathematics "for his initiation of the field of global differential geometry and his continued leadership of the field, resulting in beautiful developments that are at the center of contemporary mathematics, with deep connections to topology, algebra and analysis, in short, to all major branches of mathematics of the last sixty years."

The award committee's essay on the prize writes as follows:

Shiing-Shen Chern is the foremost geometer of our time. His mathematical work, centering on geometry and spanning almost seven decades, has shaped large areas of modern mathematics. More than any other mathematician he defined the subject of global differential geometry, one of the central areas in contemporary mathematics. The depth and originality of his mathematical insight is seen in the many basic concepts of modern mathematics to which his name is attached: Chern classes, the Chern-Weil map, the Chern connection, the Bott-Chern forms, Chern-Moser invariants, and the Chern-Simons invariants.

Chern's talent became apparent early. He completed his undergraduate work at Nankai University and his Master of Science degree work at Tsinghua University. In the 1930s he was sent to Europe, where he studied with Wilhelm Blaschke in Hamburg and Elie Cartan in Paris. Guided by their influence he wrote, in effect, two theses: one on web geometry and one on the differential invariants of a third-order ordinary differential equation. Both works were published and remain of interest today.

Chern then returned to China to teach at Tsinghua University which during the war had moved to Kunming in southwestern China. After several years in Kunming he made his way out of wartime China, and by a circuitous route via Africa managed to reach the United States. There Chern began a highly productive stay at the Institute for Advanced Study in Princeton at the invitation of Oswald Veblen and H. Weyl. It was during this period that he gave the first intrinsic proof of the general Gauss-Bonnet formula. In hindsight, one may say that from this proof evolved many fundamental concepts in topology, such as transgression in sphere bundles introduced from a differential geometric perspective. In addition, he began the monumental work that introduced Chern classes and, as a by-product, initiated the subject of Hermitian differential geometry. This work brought to the fore the relationship between differential geometry and topology; it also opened fertile new areas for other mathematicians and remains of central importance to this day.

He completed his work on the Chern classes during a brief return to China after the war. Then he moved to the University of Chicago, where together with Weil and others, he helped to establish one of the leading mathematics departments in the world. During this period his work arrived at the center of mathematical life. Through it and his influence on colleagues, Chern guided the field of differential geometry and led it into interaction with essentially all aspects of geometry, including topology, algebraic geometry, integral geometry, complex geometry, exterior differential systems, global analysis, and partial differential equations.

Chern's contributions typically begin when he engages a specific problem; then, through his geometric insight and computational mastery, he brings it to resolution;

eventually his work has been seen to open up productive new areas for other mathematicians to develop. This pattern, which continues to this day, is an extension of the classic tradition of Cartan, supplemented by a deep and far-reaching global perspective.

Two examples, one specific and one general, may serve to illustrate the continuing influence of Chern's mathematical work in the scientific community. One that is of great current interest derives from the Chern-Simons invariants, which inform both theoretical physics and three-dimensional topology. Another is Chern's recognition of the special role that a complex structure plays in differential geometry. Examples of this are evident throughout his work, including the introduction of the Chern classes of holomorphic vector bundles via curvature forms, the study of minimal surfaces and harmonic maps using the conformal structure, Chern's geometrization of complex function theory, and the geometry of CR-structures. Reflecting the pervasive role of the complex structures, the differential geometric properties of complex algebraic varieties are of central importance in modern theoretical physics and in number theory.

At the end of the 1950s Chern moved to the University of California at Berkeley, where he was a professor of mathematics and, in 1980, became founding director of the MSRI. A few years later, he also established a mathematics institute at Nankai University where he had earlier received his undergraduate education. He continued his professional life in Berkeley until five years ago, when he moved to reside at Nankai. Throughout these times he has remained mathematically active, most recently initiating a revitalization of the subject of Finsler geometry.

At Berkeley, as in all his positions of leadership, the influence of Chern as a practitioner of mathematics was matched by his warmth and skill as a teacher and leader. He was always an attentive and generous mentor to younger people. When one of us (Griffiths) first went to Princeton as a graduate student, his adviser sent him to Berkeley in the summer of 1961. Upon his arrival Chern immediately invited him to lunch, initiating a friendship and professional collaboration that continues through the present.

As a welcoming and collaborative friend, Chern likes nothing better than to gather people of all ages and perspectives to talk about mathematics. He has taken special pleasure in advancing the professional lives of those who have had the opportunity to work with him. He has always been among the first to understand the importance of the work of his colleagues and to bring it to the attention of the wider mathematical community. Chern's former students now populate the mathematics departments of major universities across the US, and his influence is, and has always been, widely felt in China.

The first Shaw Prize is awarded to Professor Chern in recognition of his singular contributions to, and influence on, the mathematical life of our time.

November 2, 2014: The Committee on Small Body Nomenclature of the International Astronomical Union approved the naming of asteroid 1998 CS2 as Chern in recognition of his contributions to humanity.

Chern Medal by the International Mathematical Union

On June 1, 2009, the International Mathematical Union officially announced the establishment of a new prestigious mathematics award—the Chern Medal: "The Chern Medal Award is established in memory of the outstanding mathematician Shiing-Shen Chern (1911, Jiaxing, China–2004, Tianjin, China). Professor Chern devoted his life to mathematics, both in active research and education, and in nurturing the field whenever the opportunity arose. He obtained fundamental results in all the major aspects of modern geometry, and founded the area of global differential geometry. Chern's work exhibited keen aesthetic tastes in his selection of problems and in his breadth exemplified the interconnectedness of modern geometry and all of its aspects."

Comments on Shiing-Shen Chern by Renowned Mathematicians

The renowned French mathematician and leader of the Bourbaki school, André Weil, said[1]: "Without the geometric intuition of Cartan, Hopf, Chern, and a few others, mathematics in this century would never have made such remarkable progress. I firmly believe that as long as mathematics continues to develop, it will always need mathematicians of their caliber."

German mathematician Peter Dombrowski (1928–) summarized the centennial history of differential geometry, saying[2]: "Overall, the ideas and work of the two geometers, Elie Cartan and Shiing-Shen Chern, have left a profound mark on the course of differential geometry during this period. Their papers reflect the development of differential geometry in many different branches."

Another renowned German mathematician, Friedrich Hirzebruch (1927–2012), a recipient of the Wolf Prize and the Lobachevsky Prize, wrote a long memorial to his mentor and friend Shiing-Shen Chern, stating[3]: "Shiing-Shen Chern was one of the greatest mathematicians of the 20th century, and an extraordinary person."

Chern's student, the renowned Chinese mathematician, Fields Medalist, and Wolf Prize laureate, Shing-tung Yau, shared[4]: "I am honored to have studied under a great mathematician. Professor Chern has had a profound impact on my academic career, both in mathematics and personal development. Looking back at the history of differential geometry, I believe Elie Cartan is the grandfather of differential geometry, and Chern the father of modern differential geometry. Together, they creat-

1 André Weil, "My Friend—the Geometer Shiing-Shen Chern," in Zhang Dianzhou and Wang Shanping, ed., *Collected Works of Shiing-Shen Chern* (Shanghai: East China Normal University Press, 2002), pp. 357–360.

2 Peter Dombrowski, *Differentialgeometrie in G. Fischer u. a. Ein Jahrhundert Mathematik* (Vieweg, 1990).

3 Friedrich Hirzebruch, and Udo Simmon, "Nachruf auf Shiing-Shen Chern," *Jahresbericht der DMV* 108, no. 8 (2006): pp. 197–217

4 Shing-tung Yau, "Shiing-Shen Chern's Contributions to Geometry," in Shing-tung Yau, Yang Le, and Ji Lizhen, eds., *Shiing-Shen Chern and the Development of Geometry* (Beijing: Higher Education Press, 2001).

ed a wonderful and rich discipline that has influenced every branch of mathematics and physics."

Chern's early student, the famous Chinese mathematician Wu Wenjun, said[5]: "Mr. Chern was not only a towering figure in mathematics but also made tremendous contributions to raising Chinese mathematics to the world stage. He successively hosted and founded two mathematical research institutes in China, training a large number of outstanding young mathematicians. This enabled Chinese mathematics to stand on par with the West, compete with the best, and created the conditions for China's mathematics to rise from a great power to a strong power by the middle of the 21st century."

The renowned Chinese mathematician and academician of the Chinese Academy of Sciences, Jiang Boju, said[6]: "Mr. Chern, with his unique charm, guided and promoted unity and progress in the mathematical community. He has always been loved by the mathematical community, not only because of his outstanding contributions to mathematics, making him a world-renowned mathematical master and the father of modern differential geometry, which may seem a bit unreachable, but also because he was a guide among us, always considering and supporting the development of mathematics, guiding and caring for us. Mr. Chern was the spiritual leader of China's mathematical reform and opening up."

---

5 Wu Wenjun and Ge Molin, preface to *Shiing-Shen Chern and Chinese Mathematics* (Tianjin: Nankai University Press, 2007).

6 Wu Wenjun and Ge Molin, *Shiing-Shen Chern and Chinese Mathematics* (Tianjin: Nankai University Press, 2007), pp. 116–117.

ABOUT THE AUTHOR

Shiing-Shen Chern (1911–2004) was a pioneering mathematician, widely regarded as the father of modern differential geometry. His work laid the foundation for major developments in global differential geometry, characteristic classes, and their applications in both mathematics and physics.

Born in China, Chern held positions at prestigious institutions including the University of Chicago, Princeton, and UC Berkeley. He was the founding director of the Mathematical Sciences Research Institute in Berkeley and a foreign member of the Chinese Academy of Sciences.

Renowned for his clarity, insight, and deep humanistic spirit, Chern was not only a mathematical giant but also a mentor and guide to generations of mathematicians.